OUTSMART THE ALGORITHM

OUTSMART THE ALGORITHM

Staying Relevant in an AI World

ROMAN KURMACHEV

Books to Hook Publishing, LLC.

CONTENTS

Let our pursuit of AI reflect the best of our human values:
Compassion, creativity, and the unending search for truth and
meaning.
In our quest to create intelligence and augment our lives, may we never
lose sight of the heart.

CHAPTER 1

INTRODUCTION

Hey awesome humans, let me lay it out for you straight - we're living in some legitimately crazy times right now. And no, I'm not talking about the latest TikTok dance craze or whatever wild meme is currently breaking the internet. I'm talking about something way bigger, way more transformative, and honestly, way more exciting: the rise of artificial intelligence.

Now, I know what some of you might be thinking. "Oh great, another doomsday prophet preaching about the robot apocalypse." But hold up, hear me out. Because while it's true that AI is shaking up just about everything we thought we knew about technology, work, and even what it means to be human, I'm here to tell you that this wild ride we're on is less of a dystopian nightmare and more of an epic adventure full of incredible opportunities - if we're bold enough to seize them.

Picture this: you're part of the first generation that gets to grow up alongside AI from day one. You're not just passive observers watching this technological tsunami crash over the world - you're the surfers riding the crest of the wave, the pioneers charting new territories, the visionaries reimagining what's possible when we combine the raw processing power of machines with the ingenuity and creativity of the human spirit.

I mean, just look at some of the mind-bending breakthroughs that are already happening thanks to AI. In healthcare, doctors are using machine learning algorithms to detect diseases like cancer earlier and more accurately than ever before. In transportation, self-driving cars are poised to revolutionize the way we get around and make our roads safer. And in fields like art and music, AI is helping to push the boundaries of what's possible, generating new sounds, styles, and masterpieces that are blowing our puny human minds.

But here's the thing - the story of AI isn't just about the incredible things that machines can do. It's about the even more incredible things that we can do when we work hand in hand with our artificially intelligent friends. That's right, I said friends. Because contrary to what some of the doomsayers might have you believe, the robots aren't coming to take our jobs and render us all obsolete. They're coming to be our partners, our collaborators, our trusty sidekicks in the grand adventure of building a better world.

And that's precisely what this book is all about - giving you the tools, frameworks, and mindsets you need to thrive in the age of AI, not by running away from the machines, but by running towards them with open arms. Throughout these pages, we'll be exploring some of the most cutting-edge research and real-world examples of how AI is transforming industries, skills, and ways of working. But more importantly, we'll be diving deep into what makes us humans so gosh-darn special and how we can leverage our unique superpowers to stay ahead of the curve.

We'll talk about the power of creativity and how to cultivate a mind that's always bursting with fresh ideas and possibilities. We'll explore the importance of emotional intelligence and how to build the kinds of deep, meaningful relationships that are the foundation of any successful career or endeavor. And of course, we'll dig into the secret sauce of critical thinking and problem-solving - the ability to look at complex challenges from multiple angles and dream up solutions that are nothing short of genius.

But we won't stop there. We'll also roll up our sleeves and get practical about how to build your personal brand in the age of AI, showcasing the skills and talents that make you a one-of-a-kind (and non-automatable) asset in any field. We'll explore the art of networking and collaboration, sharing tips for building a vibrant community of fellow humans and AI allies who can help you achieve your wildest dreams. And of course, we'll dive into the all-important topic of ethics and responsibility in the development and use of AI, because with great technological power comes...well, you know the rest.

Sounds pretty awesome, right? But don't just take my word for it. Did you know that some of the greatest minds in history have been fascinated by the idea of intelligent machines? Way back in the 19th century, Ada Lovelace, the world's first computer programmer, predicted that one day machines would be able to compose music, create art, and even function as scientific instruments - all while working side by side with their human counterparts. And in the 1950s, legendary mathematician Alan Turing proposed a test (now known as the Turing Test) to determine whether a machine could exhibit intelligent behavior indistinguishable from that of a human.

Fast forward to today, and we're seeing those visionary predictions come to life in ways that are nothing short of mind-blowing. But here's the plot twist: the story of AI isn't just about the machines - it's about us. It's about the incredible things that we're capable of when we're given the tools to amplify our intelligence, unleash our creativity, and dream bigger than ever before. And that, my friends, is a tale for the ages.

The AI age isn't approaching - it's already here, unfolding and accelerating by the day. We get to decide whether it becomes a dark dystopia or a launchpad for our species to achieve a whole new level of flourishing.

Brace yourself, because this wild ride is just getting started. Let's create something extraordinary together

1.1. From AI Hero To AI Zero

Let's start with a little story time, shall we? Because this book didn't just materialize out of thin air - it came from a very real, very personal experience that kickstarted my whole journey into the AI revolution.

Picture it: a few years ago, I was just your average tech bro, clocking in at one of those big-name companies that basically owns half the internet. My role? Fighting the good fight as a content moderator and analyst, training up AI systems to sniff out all the sketchy stuff polluting the online world.

I know, I know - not exactly the most glamorous gig. But somebody had to be on fraud and toxicity patrol, you know. Keeping the internet a relatively chill, family-friendly place and all that. Me and my scrappy team of virtual bouncers were proud to take on that challenge.

Except...our glory days didn't last too long before the whole operation went AI-first in a major way. One morning, the higher-ups basically rolled through our monitors like: "Hate to break it to y'all, but these machine learning models you've been training? Yeah, they're kinda making your roles redundant as hell."

Well, they didn't exactly say that because they use that superficial corporate talk but the real meaning of their message was clear as day. We were being put out to pasture by our own goddamn AI protégés in the name of corporate efficiencies or whatever. Cue the existential spiraling about where this whole automation wave was headed.

At first, I'll admit it stung like a cheating ex. All those late nights curating datasets, tuning models, coaching these AI newbies how to spot hate speech and misinformation...just to have them decisively whup our human capabilities and snatch our jobs right from under us.

It felt downright cold, you know? Here I was thinking we humans were steering the automation revolution from the driver's seat. The next thing I knew, we were the ones being automated out, left wondering what our relevant skills even were anymore.

But after moping around for a bit, something inside me shifted. This disruption wasn't just an inevitable change of the Robots Will Rise variety. It was a wake-up call. A chance to evolve and stop fighting the AI tide. To get upstream of the massive rapids coming our way and ride them like a pro surfer instead of drowning helplessly in the whitewash.

So I made a pact with myself: Instead of fearing or hating the machines taking our jobs, why not focus on upskilling and developing the uniquely human skills no robot could touch? Things like creativity, emotional intelligence, persuasive communication - all those traits that make our species special.

From that mindset pivot, doors just kept flying open for me. I embarked on this crazy upskilling journey, taking coding bootcamps, studying product strategy, and devouring every AI and future-of-work book I could find. And soon enough, a new path emerged where I wasn't just surviving the AI wave, but thriving in it.

That's where the mission behind this book was born. If I could go from deer-in-headlights mode to becoming an irreplaceable human-AI hybrid, I figured I could share that same perspective shift and knowledge with all of you. To help create a whole army of AI-proof personal brands and future force skill sets.

Because believe me, I get how daunting this AI renaissance feels when you're stuck in that fear-of-obsolescence loop. Having lived through the career displacement myself, I didn't just read about the challenges in some white papers - I was faced with having to upskill or risk getting straight-up robotized out of relevance.

But by getting intentional about cultivating my unique human strengths and situating myself as a strategic partner to machine capabilities rather than a linear competitor, I cracked the cheat code. And I'm dead set on spreading those same countermeasures to all you brilliant humans before the AI wave hits your industry or role.

So consider this book your tactical guide to thriving through the oncoming tsunami of automation and AI disruption heading our way. By the time we're done, you'll have all the context and tools to craft a future-proof identity centered on creativity, cognitive skills, and human-AI synergy. One that lets you soar to entirely new altitudes fueled by robotic acceleration.

No promises it'll be an easy journey. Change never is. But I can guarantee one thing: With some hard work and invested introspection from you, you'll emerge from these pages with the kind of evolutionary skills upgrade that global robocalypse can't render obsolete.

The machines are getting smarter every day. Now it's time for us fallible flesh beings to level up too. No pressure, though, we got this

1.2. The AI Awakening

Alright, let's talk about this AI awakening that's been going down in recent years. Because trust me, the rapid progress we've seen in artificial intelligence isn't just empty hype - it's the real deal. We're witnessing something that was previously confined to the realms of science fiction suddenly become reality right before our eyes.

For instance, DeepMind's AlphaGo program defeated Lee Sedol, one of the world's top Go players, back in 2016. For those not familiar, Go is this ancient Chinese board game that's famous for being crazy complex, with more potential positions than atoms in the observable universe. It was always considered way too nuanced and intuitive for an AI to master anytime soon. Until AlphaGo went beast mode, using novel machine-learning techniques to study immense datasets of human Go moves and develop frighteningly superb intuition. Lee Sedol was genuinely shook after losing multiple matches, admitting the AI's brilliant moves were like nothing he'd ever encountered from a human player.

That was a massive milestone, showing AI could now tackle extremely complicated cognitive tasks that were once thought exclusive to the human mind. But it was just a teaser for what was to come.

In 2020, OpenAI dropped their freakishly capable language model called GPT-3 and it low-key broke the internet. We're talking about an AI that can generate extremely coherent, contextual writing on pretty much any topic after studying a mountain of digital text data. From brilliant essays and poetry to snappy ad copy and even coding, this thing seemed to have a preternatural fluency with human language. Instant weekly posts hit debating whether GPT-3 had achieved some sort of artificial general intelligence (AGI) breakthrough.

Of course, the hype reached something of a fever pitch when OpenAI's wildly impressive ChatGPT bot went viral in late 2022. Suddenly, everyone and their mom were shooting the breeze with this eloquent AI, having full-on conversations and getting assistance on everything from homework assignments to self-help therapy. It was a turning point where the public got this visceral experience of how advanced and multi-talented modern AI has become at mimicking human intelligence.

Now unlike with DeepMind's games and OpenAI's language models, the technical details behind ChatGPT aren't fully known. But we do know it leverages transformers, a powerful neural network architecture that crunches gigantic datasets to generate highly contextualized outputs like human-like responses. And it's prompted massive debates around the implications of such fluent conversational AI for education, creative work, and even existential risk if the technology is pushed to extremes.

Zoom out from the headline-grabbing demos, and you can see this renaissance in AI capabilities is being turbocharged by the same forces that have catapulted technologies like smartphones and laptops into the modern era. Relentless growth in computational power following Moore's Law, the proliferation of big data from digital sources, and increasingly sophisticated machine learning models and algorithms are a triforce allowing AI to tackle increasingly complex cognitive tasks.

For historical context, consider that the AI techniques fueling recent breakthroughs like deep learning and reinforcement learning have their roots in decades-old research from brilliant pioneers like Geoffrey Hinton, David Rumelhart, and others. Their work on neural networks and biologically-inspired processing architectures laid the foundations. But it was only with recent exponential increases in data and processing power that these techniques could truly be unleashed and achieve their transformative potential.

What was once theoretical sci-fi banter about superintelligent AI is now a reality rapidly reshaping our world. That said, we're likely just glimpsing the tip of the iceberg in terms of what AI will be capable of in the coming years and decades as the tech continues its exponential leap forward. Buckle up, because the awakening has only just begun.

1.3. Invasion Of The Job Disruptors

Now let's focus on something that's probably been creeping into your minds - the notion that AI is gunning for our jobs. Because let's be real, you don't have to browse r/futurology on Reddit for long before you see wild headlines about robots making humans obsolete across every industry.

I'm not here to be an AI doomsdayer or shut down those concerns entirely. The reality is, that AI's disruptive potential is very much real and already starting to make waves in how work gets done across multiple sectors. We're in the midst of an invasion, but one that could be more of an opportunity than an outright job apocalypse if we play our cards right.

To see what I mean, let's take a look at some of the areas where AI-powered automation is already making its presence felt, replacing some human roles while augmenting others in wildly productive ways.

In healthcare, we're seeing AI diagnostic tools that can detect diseases like cancers and eye disorders way faster than human radiologists, while reducing costly errors. Dozens of FDA-approved AI algorithms are being deployed to complement doctor's expertise by rapidly processing medical images and data. No one's saying human physicians are obsolete, but their workflows are being streamlined big time.

Similar efficiency boosts are happening in fields like law and finance. Contract review used to be an incredibly tedious process taking armies of human lawyers and paralegals to pore over documents line-by-line. But now AI software can ingest those contracts, understand their provisions, and flag potential issues at lightning speed. Wall Street is seeing robo-advisors provide portfolio management and tailored advice based on advanced data analytics far beyond a human analyst's capacity.

Even uber-human domains like creative arts aren't immune to the AI job disruptors. We're entering an era where AI can code its own video games, generate photorealistic digital art, compose songs, or conjure up compelling ad copy and social media content. Human creatives may not be displaced entirely, but they'll likely need to rethink their roles and skillsets to thrive.

According to one widely-cited study by researchers at Oxford and Yale, around 47% of current U.S. jobs could be automated in the coming decades by burgeoning AI and robotic capabilities. Roles like truck drivers, cashiers, telemarketers, and data entry clerks top the "cull" lists for employment disruption.

But here's the key - that same study estimated nearly an equal amount of fresh, previously unthought-of jobs would be created in an AI-driven economy too. Things we can't even conceive of yet, which will place a premium on those uniquely human assets like creativity, emotional intelligence, and strategic decision-making that are core competitive advantages over machines.

So yes, the invasion of AI job disruptors is very much real and accelerating quickly across industries. But much like how mechanized factories didn't permanently freeze the human workforce, this robotic renaissance too will reshuffle the deck rather than checkmate humanity entirely. Smart humans who can deftly integrate with AI will thrive, while those insisting on swimming against the current may, unfortunately, find themselves submerged.

That's why going forward we all need to get intentional about retreating from activities and roles that can be optimized by narrow AI. Focus instead on cultivating skills that maximize your distinctly human value-add. Let the robots crunch numbers and scour databases. You, brilliant human, bring the spark of creativity, interpersonal dexterity, and ethical judgment that gives those AI capabilities context and purpose.

When you look back at previous waves of technological automation, from the Industrial Revolution up through the computer age, humanity didn't go extinct - we elevated. We allowed machines to handle the menial physical labor so we could lean into more cognitive, strategic work. This AI renaissance is simply the next phase of that progression, where we offload data drudgery to cyber-intelligences and embrace roles that allow our unique consciousness to soar.

The job disruptions are here and they're only going to accelerate. But with intentional adaptation and the right human-AI symbiosis, we may be invading a whole new stratum of opportunity. Kinda sounds exciting when you think about it that way, doesn't it?

1.4. Humans Need Not Apply?

I believe now is a good place to talk about the big, hairy elephant in the room when it comes to AI's epic rise - the potential for human jobs and roles to become straight-up obsolete. Because, frankly speaking, when we glimpse the dizzying pace of progress across AI capabilities like perception, data processing, decision-making, and even physical automation, it begs the million-dollar question: Are we all basically training our replacements here? Will AI eventually automate every conceivable human task to the point where we flesh-and-blood workers get ousted from employment altogether, sort of like in a post-apocalyptic thriller?

Now before you spiral into an existential crisis, take a deep breath. Yes, the technical potential for AI to encroach upon virtually every career domain is very much real based on the research and my experience. But becoming a modern-day Luddite smashing meta-machines ain't the solution either. We need to get pragmatic about the roles most susceptible to displacement so we can adapt accordingly.

Based on the studies by the Brookings Institute, McKinsey, and others, careers centered around routine physical work or data processing stand among the most immediate targets for AI disruption. We're talking jobs like truck drivers, data entry clerks, telemarketers, proofreaders, and the like. Roles that consist primarily of tightly defined, highly repetitive tasks are already being optimized by narrow AI proficient at parsing structured data and executing codified rules.

Take manual labor like operating a forklift or long-haul trucking for instance. Self-driving AI guided by sensors and cameras can increasingly handle those repetitive physical motions with more accuracy and infinitely less downtime than a human operator. Sure, edge cases and technical limitations mean we're not past the finish line yet. But the core abilities of computerized perception and robotic control are advancing at whiplash speed.

It's a similar script for data-oriented roles that were once thought to require human intelligence and discretion. But with techniques like robotic process automation (RPA), AI can now reliably ingest massive datasets, extract key details, and make nuanced decisions without getting fatigued or making silly mistakes. No wonder data entry specialists and clerk positions are being supplanted by algorithms that can crunch through exponentially more inputs at a fraction of the cost.

Even once highly-skilled roles like radiologists, financial analysts, and paralegals aren't immune. Sure, they require extensive training and credentials. But increasingly, the core functions of rapidly processing images, spreadsheets, and legal documents to extract insights or flag issues are activities prime for AI optimization. A doctor augmented by AI triage tools, an analyst flanked by automated data models - that's becoming the new normal across knowledge work.

Now does that mean the human professionals in those roles are totally obsolete? Not necessarily - at least, not yet. Their expert judgment and high-level strategic guidance complementing the AI's rapid data synthesis will likely still be valued for the foreseeable future. However, the repetitive cognitive workload that once was the bulk of those roles is indeed ripe for automation disruption.

History offers some context too. If we revisit previous industrial revolutions, we can see how technological disruption demolished certain roles before elevating entirely new labor categories. Scores of manual farmhands and textile workers saw their trades evaporate due to mechanized agriculture and manufactories. However, the ensuing economic booms raised unprecedented new employment opportunities for accountants, managers, technicians, and machine operators.

So will the AI uprising robotify every conceivable role in perpetuity until humanity is destitute and purposeless? Probably not, based on our ability to adapt and evolve in past transformations. But will narrow domains of well-defined work get gobbled up by algorithmic disruptors faster than we can sprout new jobs requiring flexible intelligence? You betcha - the history of progress is littered with casualties unable to reskill.

That's why it's crucial that we proactively cultivate intrinsically human skills like creativity, emotional intelligence and strategic decision-making which are extremely challenging to automate. And yep, some roles heavily dependent on rules and rigid processes may indeed face a grim projection of becoming obsolete before our eyes.

But that harshworked carpenter from the 1800s twisting metal spokes all day probably felt pretty obsolete too when machine disruption loomed...right until they reskilled and got elevated into a new stratum of more cognitive, engaging, infinitely scalable work. We've adapted before, and we can adapt again.

1.5. AI Supremacy Nears

How about now we take this conversation about AI disrupting human jobs to the maxed-out, most mind-bending level possible? Because up until now, we've really just been talking about relatively narrow applications of AI automating specific tasks or roles.

But what happens when we start contemplating AI systems that can match general human-level intelligence across multiple domains? I'm talking about the hypothetical "Artificial General Intelligence" (AGI) that's been a focal point of so much sci-fi speculation and futurist debate.

See, current AI like ChatGPT or AlphaGo - as impressive as they are - are still relatively specialized systems. They crunch massive data sets and excel at their designed focus like natural language processing or game strategy. But working outside those areas of training, they're about as clueless as a bricklayer trying to perform open heart surgery.

AGI, in theory, would be a generalized form of AI that can learn, reason, and transfer knowledge across multiple disciplines at a level equaling or surpassing human performance. It could dynamically take on new skills, roles, and solve open-ended problems with the type of flexible, general intelligence we effortlessly employ every day.

Now to be clear - we are still years if not decades away from actually cracking AGI. The challenges of replicating a human mind's versatility and integrating various narrow AI capabilities into a unified general intelligence are immense.

But many experts believe it's an inevitability rather than a possibility. And if or when it happens, it could represent an existential disruption to human labor and cognition itself on a scale we've never contemplated before.

Think about it - if you had an AI assistant that could understand complex contexts as well as any person, learn and update its skills dynamically, bring creativity and emotional intelligence to problems, and apply that general knowledge to any domain or task...what roles would it NOT be able to take over?

We're talking about an AGI entity that could legitimately perform open heart surgery one day, then pivot to providing psychotherapy counseling, before spending an evening ghostwriting award-winning novels or composing euphoric symphonies. It sounds ludicrous, but that's essentially the implication of replicating human-level general intelligence.

At that point, would any human profession, no matter how technical, creative, or uniquely anthropic, be insulated from potential disruption or augmentation by AGI systems? Doctors, engineers, CEOs, policymakers, entertainers - none may be impervious to an intelligence explosion that leaves biological human minds in the dust.

Now, to be clear, we have no consensus timeline for if or when this AGI reality could materialize. The AI existential risk community headed by folks like Nick Bostrom sees it as an imminent x-risk priority to be dealt with proactively. More measured voices expect a very gradual progression where AGI coexists with humans, not a single "lift-off" event.

Either way, the sheer extent of roles and labor categories an unconstrained AGI could encroach upon is what makes this the ultimate stakes for ensuring humans cultivate skills resistant to AI automation. Because once artificial general intelligence matches our versatility, what uniquely human skills or activities would remain?

Creativity? Machines already compose music and paint photorealistic art. Emotional intelligence? AI chatbots can provide psychological counseling. Strategic planning? Algorithms beat grandmasters at complex games like Go and chess.

At an existential level, we're confronted with the imperative to double down on exploring what facets of human cognition - our consciousness, our sense of purpose and ethics, our transcendent capacity for continuous self-evolution - may be the final firewall protecting roles fundamentally restricted to Homo sapiens alone.

It's a lofty, mind-warping conversation to be sure. But one we'd be prudent to engage with now rather than be reactively forced into when AGI is a clear and present disruptor to the human experience itself. The AI age is already redefining so much about society and work. But the full supremacy of AGI may be the greatest transition we've ever confronted.

CHAPTER 2

UNDERSTANDING AI'S CAPABILITIES

We've covered how AI is rapidly getting better at automating all sorts of tasks and roles that were once thought to be safely in the human domain. But let's take a step back and really try to understand the fundamental capabilities driving this AI renaissance.

Because the more you dive into the different techniques and architectures powering modern artificial intelligence, the more you realize just how mind-bendingly advanced this technology has become in a remarkably short period of time. We're talking about systems that can quite literally perceive, learn, reason and make decisions in ways that were purely science fiction just a decade or two ago.

In this chapter, we're going to be breaking down the core AI disciplines and methods that are making this happen. We'll start by unpacking the big daddy - machine learning. This is the fundamental technique allowing computers to crunch through massive datasets, identify intricate patterns, and essentially program themselves to make smart decisions and predictions, all without needing explicit lines of code from human engineers.

From there, we'll zoom in on deep learning - the specific subset of machine learning that's turbocharging AI capabilities by imitating the biological neural networks in our own gray matter. We'll explore the innovative layered architectures like convolutional and recurrent neural nets that allow deep learning models to extract profound insights directly from raw data inputs like images, speech and natural language.

With that neural net backbone established, we'll take a tour through AI's rapidly advancing perception skills across vision, speech and language domains. We'll witness how systems can now classify objects, read facial expressions, transcribe speech and understand natural language queries with scary levels of accuracy. Smart cameras, digital assistants and other mind-blowing AI apps will illustrate the implications.

But as you'll see, modern AI's abilities extend far beyond just machine perception at this point. We'll dive into how techniques like reinforcement learning are allowing AI to develop higher cognitive skills for strategic decision-making, reasoning and problem-solving. We'll even peek around the corner at the race towards artificial general intelligence (AGI) that could one day match human-level intelligence across multiple domains.

Now look, I'm not going to pretend like unpacking AI's technical capabilities is easy breezy material. There's some hairy math and coding concepts involved for sure. But I'll be breaking it down in the most understandable way possible, with plenty of relatable examples and metaphors.

Because at the end of the day, wrapping our heads around what AI can actually do is crucial for being able to think critically about its implications for society and our future. How can we have productive conversations about AI's impact on jobs, ethics and perhaps even existential risks to humanity without first understanding the scope of its current and projected abilities?

This chapter is all about arming you with that foundational knowledge. By learning the key disciplines and architectures under the hood, you'll be able to build intuitions around AI's strengths, limitations and trajectories in a way that transcends just treating it as a buzzword or hyped-up fantasy.

2.1. Machine Learning: The Brain Of AI

To start, let's break down the core concept driving this whole AI boom - machine learning. Because believe it or not, the reason why AI has gone from hokey sci-fi fantasy to world-changing reality has everything to do with giving computers a brain-like ability to learn.

See, old-school programming was all about meticulously hard-coding instructions and rules into software, line-by-line, to get computers to complete specific tasks exactly as defined by human developers. But with machine learning, we've cracked a way to essentially program computers to program themselves based on exposure to data.

Think about it like this - when you were first learning to recognize different animals as a kid, your parents didn't sit you down and rattle off a list of precise rules distinguishing dogs from cats. You looked at thousands of examples, your brain picked up on the underlying patterns like "four legs, furry, tail" and over time, you learned to reliably identify dogs, cats, and every other creature by internalizing those patterns from the training data.

Well, machine learning is kinda like recreating that human learning process, but for computer software instead of squishy brains. We feed these AI systems massive datasets, and through various techniques, they're able to study those examples and develop their own internal mathematical models to make predictions or decisions without human rules being explicitly coded.

The simplest and most common application is supervised learning, where we provide label-led training data, essentially telling the AI "This is a dog, this is a cat." The system spots patterns in the data, creates a model mapping inputs to desired outputs, and can then accurately classify new examples it has never seen before.

Unsupervised learning is the true "let the machine figure it out" approach. We just chuck a mountain of raw, unlabeled data at the AI and it has to identify inherent patterns and groupings all on its own with no guidance. Think customer segmentation for marketing campaigns based on purchasing behaviors or grouping astronomical data to dis-cover new stellar systems.

Then you've got reinforcement learning which is like that dog treat reward technique on steroids. We define a goal, like winning a chess game, and have the AI take actions to try to achieve it through plenty of trial-and-error iterations. Each "good" action gets a reward signal that reinforces the ideal strategy, without any human input on how to actually play.

What ties all these together is that instead of us rigidly coding decision rules and logic flows, the machine derives its own nuanced reasoning models by poring over data. Modern shopping recommen-dations, spam filtering, and stock predictions - all driven by machine learning algorithms that picked up their own decision strategies.

Now of course, saying "we show the AI data and it magically learns" glosses over the gnarly math and processes like neural networks under the hood. And believe me, we'll peel back the layers of that onion soon enough.

But just grasping the general concept that machines can ingest data and derive their own decision models, without needing human programmers to spoon-feed every instruction, is so powerful. It's what has allowed AI to rapidly progress in ways previous rules-based systems could never dream of.

And the crazy part? Machine learning has been theorized since the 1950s by the likes of pioneers like Arthur Samuel at IBM who created self-learning game programs. The techniques like neural nets we now use were proposed decades ago as rough approximations of how the human brain processes information.

The difference today is we have the computing power and we have the data - hundreds of millions of digital examples to feed these algo-rithms until they soak up the statistical patterns and start making scarily accurate predictions. Combine that capability with innovative new techniques and architectures that we'll explore more deeply, and you can start to see why AI has gone full renaissance in the last decade or so.

Machine learning isn't the be-all and end-all, but it's the core brain that has allowed us to breathe life into intelligent systems that can genuinely perceive, learn, and make decisions in environments too complex for rules-based programming.

And this is just the beginning. As we apply machine learning in more creative ways and augment it with other AI breakthroughs, we're tapping into a new era of automating not just physical labor, but cognitive labor too. The implications are going to be literally mind-blowing.

2.2. Deep Learning: Imitating The Human Mind

Alright, now that we've talked about the basics of how machine learning allows computers to essentially learn from data, it's time to dive deeper into what's really been turbocharging AI's capabilities in recent years - deep learning.

This is where things start getting both insanely powerful and lowkey freaky all at once. Because deep learning isn't just about feeding data into models - it's based on actually trying to recreate the neuron-firing computational architecture of the human brain...with software and math instead of grey matter.

See, our brains are made up of these interconnected networks of neurons that process information in multi-layered, hierarchical ways. When you look at an image of a dog, for example, your visual cortex isn't just flagging the entire dog object. Different neuron layers detect edges, shapes, colors, and textures and combine those features into higher-order representations before your conscious mind finally perceives "dog."

Well, deep learning models are structured in a similar multilayered, hierarchical way. They have an input layer receiving the raw data like pixels of an image. Then a series of hidden layers made up of intricate neural network nodes act like computational neurons, continuously transforming and combining the data into increasingly abstract representations.

The deeper the layers, the more expansive patterns across the entire data set the model can extract and refine. So by the final output layer, a deep learning system can derive richly informed predictions from data in its raw form - no need to manually engineer features like old computer vision systems.

Now these deep neural networks can take many architectural forms specialized for different data types. For image processing, models like convolutional neural nets have proven wickedly adept by ingeniously reducing an image into components like edges, shapes, and visual markers before reassembling it into high-level representations.

For understanding sequences like text or speech, recurrent neural networks (or their extended memory cousin, LSTMs) are the go-to. These models account for context in data by incorporating loops and passing data from their previous state to influence future outputs - much like how our brains process the meaning of words based on what came before in a sentence.

The result is flexible deep learning models that can derive insights directly from raw, unstructured data sources like images, video, audio, and natural language text in ways previous machine learning techniques struggled with. Want an AI system to transcribe speech, classify different objects, and understand intent in human queries? Deep neural nets have become the secret sauce.

And the applications just keep growing - from smart home assistants to self-driving cars to those freaky text-generating models like GPT-4 that can spit out remarkably coherent written content after training on incomprehensible text datasets. In many cases, these deep learning systems are rapidly surpassing human performance on specific tasks.

But don't get it twisted - deep learning isn't just replicating the human brain inside silicon. These models still rely on very contrived architectures, loss functions, and training data curated by engineers to optimize around specific tasks and objectives we define. Any general intelligence exhibited is still extremely narrow.

That said, we are steadily uncovering ways for deep learning to acquire more flexible cognitive abilities beyond just pattern matching.

Things like memory augmentation, reasoning modules, and transfer learning where knowledge generalizes across domains point to deep learning progressively approximating more aspects of human-level intelligence.

And when you consider the rate of innovation happening, with mathematical breakthroughs enabling more sophisticated neural nets while computational power exponentially grows...well, the future possibilities start to seem outrageous. Self-taught AI assistants with common sense? Autonomous scientific discoverers that leapfrog human researchers?

For now, deep learning's ability to extract rich hierarchical representations from raw data is already reshaping industry after industry. But some speculate its brain-mimicking efficiency may lead us to artificial general intelligence that could leave narrow human intellect in the dust.

No matter where it leads, one thing is clear - by ingeniously imitating the computational geometry of how our own minds function, deep learning has unlocked a new era where machines can quite literally build their own understanding of the world around them. The great AI awakening, it seems, has just begun

2.3. AI's Powerful Perception

Now let's discuss one of the areas where AI has pulled a total Usain Bolt and just blown past human-level abilities in the last few years - perception. I'm talking vision, speech, language understanding - all the ways we process raw sensory data from the world around us.

Thanks to those powerful deep learning techniques we just covered, AI systems can now quite literally see, hear, and comprehend their surroundings with scary accuracies that were simply science fiction not too long ago. And the consumer applications making use of this superhuman perception are already going full-on Minority Report all around us.

Let's start with computer vision since our human brains are heavily tuned for processing visual data. We're now at the point where AI image classification models can identify objects, scenes, activities and even subtle concepts like moods or clothing styles in digital images better than many humans can.

Those deep convolutional neural nets I mentioned? They've allowed vision AI to make sense of the billion-odd pixels in an image by progressively extracting higher-level features - edges become shapes, shapes become objects, and so on until it comprehends the overall context with eerie preciseness.

You may have experienced this capability yourself through services like Google Images search by uploading a photo. The AI can identify "dog, grass, frisbee, fetch" based solely on its raw learned perception without any text clues. Similarly, apps like Google Lens can identify objects, buildings, plants and more through your smartphone camera.

But it goes way beyond just classification too. State-of-the-art object detection models can locate and draw actual boundary boxes around every distinct object in an image in real time, handling scenarios as varied as visible NBA players to microscopic cancer cells for medical imaging.

Facial recognition? Detecting and identifying specific people in photos? Modern AI handles that will jaw-dropping accuracy by mapping and matching the unique facial geometry details humans struggle to consciously perceive.

It's got scary implications for surveillance and privacy, sure. But it's also allowing us to do incredibly useful things like automatically organize personal photo libraries, sense emotional states for mental health tracking, or prevent identity fraud.

Now take all those superhuman vision capabilities and imagine combining them with speech recognition models that can transcribe audio with near-perfect accuracy across multiple languages. Siri, Alexa, and other virtual assistants employ these speech-to-text deep learning techniques to understand our naturally spoken voice commands.

But it doesn't stop at just transcribing - natural language processing models can derive the actual intent and context within written passages or transcribed speech with shocking fluency. We're talking about AI that doesn't just robotically see "dog" and "frisbee", but understands the overall activity depicted is canine fetch playtime.

So now you've got AI assistants that can engage in back-and-forth conversations, precisely answering our questions by processing both the literal meaning and contextual subtext. It's wild to consider we interact with interfaces that cohesively combine perception across vision, speech, and language every single day now.

And this unified AI perception is turbocharging self-driving cars too - where models analyze streaming sensor data from cameras, LIDAR, and radar to make real-time decisions detecting obstacles, reading signage, and predicting movement around the vehicle.

Look, the marquee applications showing off AI's cognitive prowess like chess mastery and protein structure prediction get lots of hype. But I'd argue this ability to perceive and understand the world through raw sensory data is where AI has undergone its most startling revolution in just a few years.

Tasks like automated video understanding, surveillance analytics, and real-time transcription services that seemed unfathomable for machines not long ago are suddenly achievable and rapidly becoming ubiquitous. And this is all still just narrow computer perception - imagine what happens when we uncap AI's general intelligence exponentially.

Whether you find it awesome or terrifying, there's no denying AI's new Terminator-like abilities to see, hear, and comprehend its surroundings are quickly becoming our reality. The 21st century is turning into one wild cyborg circus already

2.4. Beyond Perception: Cognition And General Intelligence

We've covered how AI has straight-up dunked on human abilities when it comes to raw perception skills like vision and speech recognition. But the real question is - can these silicon minds actually think and reason like us carbon-based brainiacs?

The answer might surprise you. Because while AI's cognitive chops are still a bit behind its perceptual prowess, we're seeing some serious strides being made in getting computers to mimic higher-level human intelligence. I'm talking problem-solving, logical reasoning, strategic decision-making - all the mental gymnastics that separate us from the rest of the animal kingdom.

One of the key techniques pushing this cognitive progress is reinforcement learning. It's kinda like training a dog, except instead of treats, you've got an AI agent trying to maximize a predefined reward in a given environment through trial and error. The agent takes actions, sees the outcomes, and over many (many) iterations, learns an optimal strategy to achieve its goal.

This approach has led to some seriously impressive results like DeepMind's AlphaGo becoming the first computer program to defeat a world champion at the insanely complex game of Go. By playing millions of games against itself and using reinforcement learning to gradually develop its own strategic intuition, AlphaGo demonstrated a level of long-term planning and decision-making prowess that shocked even its own creators.

Since then, reinforcement learning has been applied to tasks as varied as optimizing data center cooling systems, designing new molecules for drug discovery, and even teaching robots to navigate cluttered environments with human-like agility. But it's not just about mastering specific domains - the holy grail is to create AI systems that can learn and adapt to any new challenge thrown their way.

This is where approaches like meta-learning come into play, which aims to build AI that can essentially "learn how to learn." Instead of being trained on a single task, these models are exposed to a wide variety of different problems and learn a generalized learning strategy that can be applied to novel situations. It's like giving the AI a superpower to rapidly pick up new skills and knowledge on the fly.

Another key breakthrough in pushing towards human-like cognition is the development of neuro-symbolic AI systems. The idea is to combine the pattern recognition abilities of deep learning with the structured logical reasoning of classic symbolic AI. So you've got a neural network that can perceive and extract information from raw data, feeding that into a symbolic system that can perform complex reasoning and inferences.

This hybrid approach has shown a lot of promise in tasks like visual question answering, where the AI needs to not just recognize objects in an image, but reason about their relationships and attributes to answer abstract queries. By leveraging both the perceptual powers of deep learning and the logical rigor of symbolic methods, neuro-symbolic AI is starting to display the kind of flexible intelligence that has long eluded machines.

But the grandest ambition in the quest for artificial cognition is the development of Artificial General Intelligence (AGI) - machines that can match or even surpass human intellect across any domain. We're talking about an AI that could hop from writing novels to proving mathematical theorems to composing symphonies, all while engaging in witty banter and pondering the nature of its own existence.

Now, we're still likely decades away from achieving true AGI, and there's a raging debate in the AI community about whether it's even possible with our current approaches. Some argue that just extending current deep learning methods will eventually get us there, while others think we need fundamental breakthroughs in areas like common sense reasoning and natural language understanding.

But make no mistake - the race towards AGI is very much on, with tech giants and brave startups alike pouring billions into cracking this ultimate challenge. And the potential implications are both exhilarating and terrifying. An AI system with human-level general intelligence could accelerate scientific discoveries, solve complex global challenges, and usher in a new era of abundance. But it could also pose existential risks if not developed with careful safety controls and alignment with human values.

So while we marvel at AI's incredible perceptual abilities today, keep an eye on the cognitive revolution brewing just beneath the surface. Through approaches like reinforcement learning, neuro-symbolic AI, and the pursuit of AGI, we are slowly but surely building machines that don't just see and hear like humans - but think and reason like us too.

The age of artificial intelligence may soon become the age of artificial minds. And when that happens, it'll make the Industrial Revolution look like a mere historical footnote. In any case, the future is going to be a wild ride!

2.5. Natural Language Processing (NLP): Understanding Human Language

One of the most mind-blowing and rapidly evolving areas of AI that we are about to cover is the field of Natural Language Processing, or NLP for short. I know that might sound like some kind of fancy linguistic jargon, but trust me, it's a lot cooler (and more important) than it might seem at first glance.

You see, language is at the very heart of what makes us human. It's how we communicate with each other, express our thoughts and feelings, and make sense of the world around us. And for a long time, it was something that only humans could do - after all, have you ever tried to have a deep, meaningful conversation with Siri or Alexa? Yeah, not so much.

But thanks to the incredible advancements in NLP over the past few years, machines are starting to get scary good at understanding and even generating human language. And the implications of that are nothing short of revolutionary.

So, what exactly is NLP? In a nutshell, it's all about teaching computers how to process, interpret, and analyze huge amounts of human language data, whether it's in the form of text, speech, or even social media posts. The goal is to help machines extract meaning and insights from all that messy, unstructured data in a way that's useful and action-able for humans.

And boy, have we come a long way in a short amount of time. Just a few decades ago, NLP was still in its infancy, with early chat-bots like ELIZA (developed way back in the 1960s) relying on simple pattern matching and pre-programmed responses to mimic human conversation. But fast forward to today, and we've got AI systems that can engage in remarkably coherent and contextual dialogues, thanks to advances in deep learning and neural language models.

One of the most impressive examples of this is GPT-3 (short for "Generative Pre-trained Transformer 3"), a language model developed by OpenAI that can generate human-like text on just about any topic with uncanny fluency and coherence. Trained on a massive corpus of online data, GPT-3 has been used for everything from writing articles and stories to generating code and even creating memes. It's a powerful demonstration of just how far NLP has come in terms of understanding and replicating the nuances of human language.

But NLP isn't just about generating funny tweets or writing the next great American novel. It's also being used in a wide range of practical applications that are transforming industries and changing the way we interact with technology on a daily basis.

Take chatbots, for example. If you've ever used a customer service chat window on a website or app, chances are you've interacted with an NLP-powered bot at some point. These virtual assistants use NLP techniques to understand the intent behind your queries, provide relevant information or solutions, and even handle basic transactions like booking appointments or making purchases. And as the technology continues to improve, chatbots are becoming increasingly sophisticated and human-like in their interactions.

Another big area where NLP is making waves is in sentiment analysis - the process of using algorithms to identify and extract subjective information from text data, such as opinions, attitudes, and emotions. This is incredibly valuable for businesses and organizations looking to gauge public opinion on their products, services, or brand in real-time. By analyzing social media posts, reviews, and other user-generated content, companies can get a pulse on what people are saying about them and adjust their strategies accordingly.

And let's not forget about the incredible advancements in machine translation, which is powered by NLP techniques. Gone are the days of clunky, word-for-word translations that barely made sense - today's AI-powered translation tools can handle complex grammar, idioms, and even slang with remarkable accuracy. This is breaking down language barriers and enabling people from all over the world to communicate and collaborate like never before.

But perhaps the most exciting thing about NLP is that we're still just scratching the surface of what's possible. As machines get better and better at understanding and processing human language, the potential applications are almost limitless - from more natural and intuitive voice assistants to AI-powered language learning tools to even machines that can engage in creative writing and poetry.

Of course, there are also important ethical considerations to keep in mind as NLP continues to evolve. As machines become more adept at generating human-like language, there are risks of things like misinformation, bias, and even impersonation that will need to be carefully monitored and mitigated. But overall, the field of NLP represents one of the most exciting and transformative frontiers in the world of AI, and one that has the potential to fundamentally change the way we interact with machines and with each other.

So the next time you're chatting with a customer service bot or using Google Translate to decipher a foreign language menu, take a moment to appreciate the incredible complexity and power of NLP. And if you're a budding linguist or computer scientist looking to make your mark in the field, well, there's never been a better time to dive in and start exploring the fascinating world of human-machine communication. The future of language is being written as we speak - and you have the chance to be a part of it!

CHAPTER 3

THE FUTURE OF
WORK

Alright, folks, it's time to address the 800-pound gorilla in the room whenever we start waxing poetic about the incredible capabilities of artificial intelligence. I'm talking about the very real, very pressing question on everyone's minds: what's this whole robot revolution going to mean for the future of our jobs and livelihoods?

It's a heavy topic, I know. The idea that the work we've spent years training for, the careers we've invested blood, sweat, and tears into building, could one day be taken over by a bunch of algorithms and machine learning models - it's enough to send anyone spiraling into an existential crisis.

But here's the thing - burying our heads in the sand and pretending this AI disruption isn't coming isn't doing us any favors. If we really want to thrive in this brave new world, we need to face these challenges head-on, get informed about which roles are most vulnerable, and start strategizing about how to adapt and evolve our skill sets accordingly.

That's exactly what we're going to dive into in this chapter. We'll be taking a hard look at the cold, hard data on which jobs and tasks are most likely to get automated away by AI in the near and long-term future. And trust me, the research doesn't pull any punches - from factory workers to financial analysts, the disruptive potential is very real across a wide swath of industries.

But we're not just going to be focusing on the doom and gloom here. Because the truth is, for every role that gets automated, there's also an opportunity for new, previously unthought-of jobs to emerge in the post-AI economy. We'll be exploring what some of those might look like, and how you can start positioning yourself to ride that wave rather than getting swallowed up by it.

First up, we'll be digging into the tasks and roles that are basically sitting ducks for automation because of how repetitive, routine, and rule-based they are. I'm talking data entry, telemarketing, and basic manufacturing assembly lines - the kinds of work AI is already more than capable of handling with speed and meticulous, error-free execution. It might seem obvious, but the scale of employment in these spheres means serious turbulence ahead.

From there, we'll move up the complexity ladder to roles requiring more advanced cognition, specialized knowledge, and problem-solving abilities. Think paralegals wrangling contracts, financial analysts crunching numbers, radiologists analyzing medical images. While these jobs aren't going away overnight, AI's rapid gains in parsing unstructured data and applying expert domain knowledge means the writing is increasingly on the wall for disruption.

Now I know what you might be thinking - sure, AI can crunch data and automate routine stuff, but it'll never be able to master true human creativity, right? Well, I hate to break it to you, but even those artistic and innovative abilities we cherish may not be as future-proof as we'd like to believe. With AI already composing music, generating digital art, and displaying flashes of computational creativity, no role is truly safe from the robo-renaissance.

But perhaps most concerningly, we'll explore how even the most complex, "human" roles requiring emotional intelligence, strategic vision, and socio-political savviness could one day be challenged by the rise of artificial general intelligence (AGI). Once AI reaches human-level cognitive abilities, all bets are off for which jobs remain a human birth-right vs. what the robots will come for.

Now listen, I'm not here to spark a mass panic or have you all rage-quitting your jobs to go live in the woods off the grid. Quite the contrary - my goal with this chapter is to empower you with clear-eyed, data-driven knowledge of what skills and roles are most vs. least vulnerable in the coming years so you can think proactively about your own future-proofing.

Because here's the unshakeable truth - in an increasingly AI-driven world, complacency is the biggest career killer of all. Those who bury their heads in the sand, stubbornly clinging to the status quo and banking on their jobs being untouchable, are the ones most likely to get steamrolled by the automation avalanche.

But for those of you bold enough to stare down the disruptive realities, to cultivate the human skills and competencies that will be at a premium in the new machine age - well, you're the ones who won't just survive this AI revolution but will have a once-in-a-lifetime chance to become the pioneers of a new world of work we can barely imagine yet.

So grab a coffee, get comfortable and let's dive unflinchingly into the data on our employment destiny. It won't be an easy pill to swallow, but I promise you'll emerge fired up and ready to seize the amazing opportunities ahead too. The future of work is ours to write - but only if we put in the pen to paper starting now!

3.1. Automating The Automatable

Let's kick things off by diving into the roles and tasks that are basically sitting ducks for AI automation in the near future. I'm talk-ing about jobs that are so repetitive, so routine, and so rule-based that you could practically train a monkey to do them (no offense to our primate pals).

These are the kinds of gigs where you're essentially following a script or a set of predefined steps day in and day out. There's not a lot of room for creativity, critical thinking, or dealing with curveball scenarios.

And that, my friends, is exactly what makes them prime candidates for being replaced by AI systems that can crunch through those tasks with lightning speed and meticulous accuracy.

Take data entry clerks, for example. You know, the unsung heroes of the corporate world who spend their days manually inputting endless streams of information into databases and spreadsheets. It's a tedious, mind-numbing job that's basically tailor-made for automation. I mean, why pay a human to do that when you can have an AI algorithm do it faster, cheaper, and with virtually zero errors?

The same goes for telemarketers, those poor souls who have to spend hours on end cold-calling people and reading from a script, trying to convince them to buy some product or service they probably don't need. It's a thankless job that's already being automated by AI-powered chatbots and voice assistants that can handle those conversations with inhuman patience and persistence.

Even bank tellers, who were once the friendly face of our financial transactions, are increasingly being replaced by AI-powered ATMs and mobile banking apps that can handle most routine tasks like deposits, withdrawals, and account inquiries. And let's be real - when was the last time you actually went into a physical bank branch to talk to a teller? Chances are, you've been interacting with AI this whole time without even realizing it.

But perhaps the most iconic example of jobs ripe for automation is the assembly line workers in factories and manufacturing plants. These are the folks who spend their days performing repetitive tasks like welding, painting, or putting together the same widget over and over again. It's the kind of work that's been the backbone of industrial economies for centuries, but now even that is being disrupted by AI-powered robots that can do those tasks with superhuman speed, precision, and endurance.

In fact, this idea of machines replacing human labor in factories is nothing new. If we take a little stroll down history lane, we can see that this has been happening since the early days of the Industrial Revolution in the late 18th century. That's when inventions like the spinning jenny and the power loom first started automating textile production, displacing countless skilled artisans and craftspeople in the process.

Fast forward to the early 20th century, and we had the rise of the assembly line and mass production techniques pioneered by Henry Ford and others. Suddenly, tasks that had once required skilled human labor could be broken down into simple, repeatable steps that could be performed by machines or unskilled workers. It was a seismic shift that transformed manufacturing and laid the groundwork for the modern consumer economy.

But here's the thing - as disruptive as those earlier waves of automation were, they pale in comparison to what AI is capable of today. Because now we're not just talking about machines that can perform physical tasks, but machines that can actually think, learn, and make decisions in ways that were once the exclusive domain of human cognition.

And that's why the roles and tasks we're talking about in this chapter are just the tip of the iceberg when it comes to AI-driven job displacement. According to a study by McKinsey Global Institute, up to 30% of tasks in 60% of occupations could be automated using currently available technologies. That means that even if your job isn't fully automated, there's a good chance that at least some of your daily tasks and responsibilities could be.

Now, I know that all of this might sound a bit doom and gloom, like we're all just sitting around waiting for the robot overlords to come and take our jobs. But here's the thing - while AI will undoubtedly disrupt many traditional roles and industries, it will also create new opportunities and jobs that we can't even imagine yet.

The key is to start thinking proactively about how you can position yourself to thrive in this brave new world. That means focusing on developing the kinds of skills and competencies that will be harder for AI to replicate, like creativity, empathy, strategic thinking, and adaptability. It means being open to continuous learning and upskilling throughout your career, so you can stay ahead of the curve as the job market evolves.

And who knows - maybe one day, instead of being replaced by AI, you'll be the one designing, building, and working alongside these incredible machines to create value in entirely new ways. The future is always uncertain, but one thing is clear - those who embrace the change and lean into the opportunities of AI will be the ones who thrive in the decades to come.

3.2. The Cognition Crunch

So far, we've talked about how AI is already automating a lot of the repetitive, routine tasks that have been the bread and butter of many jobs for decades. But what about the more complex, specialized roles that require years of education, training, and experience? Surely those are safe from the robot takeover, right?

Well, not so fast. Because as AI continues to advance in its ability to reason, solve problems, and develop deep domain expertise, even these so-called "knowledge work" jobs are starting to look increasingly vulnerable to automation.

Take paralegals, for example. For years, these legal professionals have been the unsung heroes of law firms, doing a lot of the heavy lifting when it comes to researching cases, reviewing contracts, and preparing documents. It's a job that requires a deep understanding of legal concepts and procedures, as well as the ability to think critically and pay attention to detail.

But now, AI is starting to encroach on that territory as well. With the ability to process vast amounts of legal data and identify relevant patterns and precedents, AI-powered legal research tools like ROSS Intelligence and Casetext are already starting to automate many of the tasks that paralegals used to do manually. And as these tools get smarter and more sophisticated, it's not hard to imagine a future where much of the paralegal's job could be done by machines.

The same goes for financial analysts, who use their expertise in accounting, economics, and data analysis to help companies and investors make informed decisions about investments and financial strategies. It's a job that requires a lot of specialized knowledge and experience, as well as the ability to think critically and make complex judgments based on a wide range of data points.

But here too, AI is starting to make inroads. With the ability to process and analyze vast amounts of financial data in real time, AI-powered investment tools like Kensho and Numerai are already starting to automate many of the tasks that financial analysts used to do manually. And as these tools get more advanced, it's not hard to imagine a future where much of the financial analyst's job could be done by machines as well.

Even highly specialized fields like radiology, which require years of medical training and expertise to interpret complex medical images, are starting to feel the heat from AI. With the ability to analyze vast amounts of medical data and identify patterns and anomalies that humans might miss, AI-powered diagnostic tools like IBM Watson Health and Google DeepMind are already starting to assist radiologists in their work. And as these tools get more accurate and reliable, it's not hard to imagine a future where much of the radiologist's job could be automated as well.

Now, I know what you might be thinking - surely there are some knowledge work jobs that are just too complex and nuanced for machines to replicate, right? Like software testing, for example - a job that requires a deep understanding of how software works, as well as the ability to think creatively and identify edge cases and potential bugs.

But even here, AI is starting to make inroads. With the ability to generate and execute complex test cases based on machine learning algorithms, AI-powered testing tools like Applitools and Testim are already starting to automate many of the tasks that software testers used to do manually. And as these tools get smarter and more sophisticated, it's not hard to imagine a future where much of the software tester's job could be done by machines as well.

Of course, this is not the first time that technology has disrupted knowledge work jobs. If we look back at history, we can see plenty of examples of how automation has transformed industries and displaced skilled workers.

Take the printing industry, for example. In the 15th century, the invention of the printing press by Johannes Gutenberg revolutionized the way that books and other written materials were produced. Suddenly, instead of relying on skilled scribes to copy texts by hand, printers could mass-produce books quickly and cheaply using movable type. This innovation led to the rise of a whole new industry, but it also displaced many of the skilled craftsmen who had previously made their living as scribes and illuminators.

Similarly, in the 20th century, the rise of digital technologies like computers and the internet transformed many knowledge work industries. Suddenly, tasks that had previously required teams of skilled workers could be automated using software and algorithms. This led to the rise of whole new industries like software development and data science, but it also displaced many traditional knowledge workers like typists, secretaries, and librarians.

So while the idea of AI automating knowledge work jobs might seem like a new and scary development, in many ways, it's just the latest chapter in a long history of technological disruption. The key is to start thinking proactively about how you can position yourself to thrive in this new world, by developing the kinds of skills and competencies that will be harder for machines to replicate.

That might mean focusing on tasks that require a high degree of creativity, empathy, and human judgment, like strategy, leadership, and innovation. Or it might mean developing a deep expertise in a particular domain and using that knowledge to work alongside AI tools in new and innovative ways.

Whatever path you choose, the important thing is to stay adaptable, curious, and open to new possibilities. Because while the future of work might be uncertain, one thing is clear - those who are able to embrace change and learn new skills will be the ones who thrive in the age of AI.

3.3. When Robots Get Creativity

Okay, we've covered how AI is starting to automate a lot of jobs that involve repetitive tasks or specialized knowledge. But what about the really creative stuff - you know, the things that make us uniquely human, like art, music, and writing? Surely that's one area where we'll always have an edge over the machines, right?

Well, I hate to be the bearer of bad news, but even in the realm of creativity, AI is starting to make some serious inroads. And while it might be hard to imagine a robot composing the next great symphony or writing the next Great American Novel, the truth is that AI is already starting to do some pretty impressive things in the creative space.

Let's start with art and design. For years, this has been seen as one of the last bastions of human creativity - after all, how could a machine possibly replicate the vision and inspiration of a great artist? But with the rise of AI-powered tools like DALL-E, Midjourney, and Stable Diffusion, even that is starting to change.

These tools use a technique called generative adversarial networks (GANs) to create incredibly realistic and imaginative images based on textual descriptions or visual prompts. Basically, the AI is trained on massive datasets of existing images and then uses that knowledge to generate new images that are similar in style or content.

And the results are pretty mind-blowing. With just a few simple prompts, these tools can create stunning landscapes, portraits, and even abstract designs that look like they were made by human artists. Some of them are so good that they've even won art competitions and been featured in galleries around the world.

Of course, this has led to a lot of hand-wringing among professional artists and designers who worry that AI might eventually put them out of a job. After all, if a machine can create art that's just as good (or better) than what a human can do, why would anyone pay for the real thing?

But it's not just visual art that's being disrupted by AI. Even in the world of music, machines are starting to make some serious waves. With tools like Google's Magenta and IBM's Watson Beat, AI is now able to compose original music in a variety of genres and styles, from classical to pop to jazz.

These tools work by analyzing massive datasets of existing music and then using that knowledge to generate new compositions that follow similar patterns and structures. And while the results might not be quite as good as what a human composer could create, they're getting better all the time.

In fact, some AI-generated music has even been used in commercial projects, like video games and movies. And with the rise of virtual influencers and digital avatars, it's not hard to imagine a future where AI-powered musicians and bands become just as popular as their human counterparts.

But perhaps the most impressive (and unsettling) area where AI is starting to make creative inroads is in the world of writing. With tools like GPT-3 and others based on large language models, AI is now able to generate incredibly coherent and engaging text on just about any topic imaginable.

These tools work by analyzing massive amounts of existing text data, like books, articles, and websites, and then using that knowledge to generate new text that follows similar patterns and structures. And while the results might not be quite as polished or insightful as what a human writer could produce, they're getting better all the time.

In fact, some AI-generated articles and stories have even been published in major news outlets and literary magazines. And with the rise of automated journalism and content creation, it's not hard to imagine a future where much of the writing we consume on a daily basis is actually produced by machines.

Now, I know what you might be thinking - this all sounds pretty dystopian, like we're headed towards a future where machines will replace human creativity entirely. But the truth is, it's not quite that simple.

For one thing, even as AI gets better at generating creative content, there will always be a place for human creativity and imagination. After all, machines can only work with the data they're given - they can't come up with truly original ideas or insights in the same way that humans can.

And for another, the rise of AI in creative fields might actually open up new opportunities for human creators to collaborate with machines in interesting and innovative ways. Just like how the invention of the camera didn't replace painters but rather gave them a new tool to work with, AI could end up being a powerful creative partner for artists, musicians, and writers.

Of course, this isn't the first time that technology has disrupted creative industries. In the early 20th century, the rise of recorded music and radio threatened the livelihoods of live musicians and performers. And in the late 20th century, the rise of digital media and the internet upended traditional publishing and media industries.

But in each case, human creativity found a way to adapt and thrive in the face of technological change. And there's no reason to think that the same won't be true in the age of AI.

So while it's true that AI is starting to make some pretty impressive inroads into creative fields, it's important to remember that creativity is ultimately a human endeavor. And as long as we continue to value and cultivate our own creative abilities, there will always be a place for human artists, musicians, and writers in the world - even if they end up working alongside some pretty impressive machines.

3.4. The Final AI Frontier

So we've talked about how AI is starting to automate a lot of jobs that involve repetitive tasks, specialized knowledge, and even creative work. But what about the really high-level stuff - you know, the jobs that require complex reasoning, emotional intelligence, and a deep understanding of human behavior and social dynamics?

Surely those are the roles that will always be reserved for humans, right? After all, how could a machine possibly replicate the nuance and intuition of a great leader or the empathy and wisdom of a skilled counselor?

Well, it's a fair question. And for a long time, many experts believed that these kinds of jobs would indeed be the last bastion of human labor in an increasingly automated world. But as AI continues to advance at a rapid pace, even that assumption is starting to be challenged.

To understand why, we need to talk about something called artificial general intelligence, or AGI for short. This is the idea of a machine that can think and reason in the same way that a human can - not just in narrow, specialized domains like playing chess or recognizing images, but across a wide range of tasks and contexts.

Now, we're still a long way off from achieving true AGI - most experts believe we're at least decades away from that level of sophistication. But the rapid progress being made in fields like machine learning, natural language processing, and cognitive computing suggests that it might not be as far-fetched as we once thought.

And if we do manage to create an AGI system that can match or exceed human-level intelligence across the board, then all bets are off when it comes to which jobs will be safe from automation.

Take managers and leaders, for example. We tend to think of these roles as requiring a high degree of emotional intelligence, strategic thinking, and people skills - all things that machines have traditionally struggled with. But with the rise of AI-powered tools like sentiment analysis and natural language processing, even these skills are starting to be replicated by machines.

Imagine an AI system that can analyze vast amounts of data on employee performance, team dynamics, and market trends, and then use that information to make strategic decisions and provide personalized coaching to individual team members. Or a virtual manager that can handle scheduling, performance evaluations, and conflict resolution with the same level of skill and efficiency as a human leader.

Of course, we're not quite there yet. But the building blocks for these kinds of systems are already starting to be developed. And as AI continues to advance, it's not hard to imagine a future where many of the tasks currently performed by human managers and leaders are automated or augmented by machines.

The same goes for roles like counselors and therapists. We tend to think of these professions as requiring a deep understanding of human behavior and emotions, as well as the ability to build trust and rapport with clients. But here too, AI is starting to make inroads.

There are already chatbots and virtual therapy apps that can provide basic mental health support and guidance to users via smartphone or computer. These tools use natural language processing and sentiment analysis to understand and respond to user input in a way that mimics human conversation and empathy.

And while they're not yet as sophisticated as human therapists, they're getting better all the time. In fact, some studies have suggested that people may actually be more likely to open up to virtual counselors than to human ones because they feel less judged or stigmatized.

But perhaps the most far-reaching impact of AGI would be on roles like policymakers and government leaders. These are the people who make decisions that affect millions of lives and shape the course of entire societies. And while we might like to think that these decisions are based on rational analysis and objective facts, the truth is that they're often influenced by a complex web of social, cultural, and political factors.

So what happens when you have an AGI system that can analyze vast amounts of data on everything from economic trends to public opinion to geopolitical risks, and then use that information to generate policy recommendations and decisions?

On one level, it's a tantalizing prospect - the idea of a perfectly rational, objective decision-maker that can optimize for the greater good without being swayed by personal biases or political pressures. But on another level, it's a deeply unsettling thought - the idea of handing over the reins of power to a machine that may not share our values or priorities.

Of course, this is all still in the realm of speculation at this point. But as AGI continues to advance, it's an issue that we'll need to grapple with sooner or later. And it's not just a question of which jobs will be automated, but of what kind of society we want to build in the age of intelligent machines.

Will we use AGI to augment and enhance human decision-making, or will we cede control entirely to the algorithms? Will we prioritize efficiency and optimization over human values and ethics, or will we find a way to integrate the two?

These are not easy questions to answer. But they're ones that we'll need to start thinking about as the final frontier of AI draws ever closer. And while it's easy to get caught up in the hype and the fear surrounding AGI, it's important to remember that ultimately, the future is still ours to shape.

3.5. Envisioning The Workplace Of The Future

Now I suggest we take a moment to peer into our crystal balls and imagine what the workplace of the future might look like in the age of AI. Picture this: you wake up in the morning, grab your coffee, and sit down at your computer to start your workday. But instead of logging into a physical office, you're connecting with your team members from all around the world in a virtual workspace powered by AI and advanced collaboration tools.

Gone are the days of long commutes, stuffy cubicles, and endless meetings that could have been an email. In the AI-driven workplace of the future, the traditional boundaries between work and life will become increasingly blurred, and the focus will shift from "where" you work to "how" you work.

But wait, isn't remote work and virtual collaboration already a thing? That's right, to an extent. The COVID-19 pandemic certainly accelerated the trend towards remote work and digital collaboration, and many companies have already embraced this new way of working. But the workplace of the future will take this to a whole new level, thanks to the power of AI and other advanced technologies.

Imagine a virtual workspace where AI-powered tools can help you manage your tasks, prioritize your workload, and even suggest new ideas and solutions based on your past work and preferences. Imagine being able to collaborate with your colleagues in real-time, no matter where in the world they are, using virtual reality and augmented reality tools that make it feel like you're in the same room. And imagine having access to a wealth of data and insights at your fingertips, thanks to AI-powered analytics and visualization tools.

But the workplace of the future won't just be about fancy tech and cool gadgets - it will also be about creating a more human-centered, employee-friendly environment that prioritizes well-being and work-life integration. With AI and automation taking over many of the repetitive and mundane tasks that used to take up so much of our time, we'll have more opportunities to focus on the things that truly matter - creativity, innovation, and meaningful connections with our colleagues and customers.

Of course, this vision of the future workplace isn't without its challenges and potential pitfalls. For one thing, the increased reliance on remote work and virtual collaboration could lead to feelings of isolation and disconnection among employees, particularly if proper support systems and communication channels aren't in place. There's also the risk of AI and automation being used to monitor and control workers in ways that feel invasive or oppressive, rather than empowering and supportive.

But these are challenges that can be overcome with careful planning, thoughtful leadership, and a commitment to putting people first. And there are already plenty of examples of companies and organizations that are paving the way towards a more human-centered, AI-driven workplace.

Take the example of GitLab, a software company that has been fully remote since its founding in 2014. GitLab has developed a comprehensive set of tools and practices to support remote work and virtual collaboration, including regular video check-ins, asynchronous communication channels, and a focus on results rather than hours worked. They've also implemented AI-powered tools to help manage projects, track progress, and even onboard new hires - all while maintaining a strong culture of transparency, trust, and employee autonomy.

Or look at the example of Unilever, the global consumer goods company, which has been experimenting with AI-powered tools to support employee well-being and work-life integration. They've developed an AI-powered coaching app called "Wellbeing4U" that provides personalized recommendations and support for employees based on their individual needs and preferences - from stress management and mindfulness exercises to nutrition and fitness advice. And they've also implemented AI-powered scheduling tools to help employees balance their work and personal commitments in a way that feels sustainable and fulfilling.

These are just a couple of examples, but they point to a larger trend towards a more human-centered, AI-driven workplace that prioritizes employee well-being and work-life integration. And as more and more companies start to embrace this vision of the future, we can expect to see even more innovative and exciting developments in the years to come.

But of course, as with any major technological or societal shift, there will be challenges and growing pains along the way. We'll need to be proactive and intentional about addressing issues like digital equity and access, data privacy and security, and the potential for AI and automation to exacerbate existing inequalities and power imbalances.

We'll also need to be mindful of the human element in all of this - the fact that no matter how advanced our technology becomes, we are still fundamentally social creatures who crave connection, meaning, and purpose in our work and lives. We can't just rely on AI and automation to solve all of our problems - we need to actively cultivate the skills and mindsets that will allow us to thrive in this new world of work, from emotional intelligence and creative problem-solving to adaptability and resilience.

But if we can do that - if we can harness the power of AI and other advanced technologies in a way that empowers and uplifts employees, rather than replacing or oppressing them - then the possibilities for the workplace of the future are truly endless. We could create a world where work is no longer a source of stress and drudgery, but a pathway to growth, fulfillment, and human flourishing.

CHAPTER 4

COMPLEMENTING AI >
FIGHTING AI

For now, let's shift gears a bit and talk about how we can actually thrive in this brave new world of artificial intelligence, rather than just fear it or try to fight it. Because if there's one thing that's become clear from our exploration of AI so far, it's that this technology is not going away anytime soon - in fact, it's only going to become more pervasive and more powerful as time goes on.

So the question becomes - how do we position ourselves to not just survive, but actually thrive in a world where AI is increasingly taking over many of the tasks and jobs that we once thought were the exclusive domain of humans?

Well, that's exactly what we're going to dive into in this chapter. And the key message I want you to take away is this: rather than seeing AI as a competitor or an adversary that we need to fight against, we need to learn how to complement and work alongside these incredible technologies in ways that leverage our uniquely human strengths and abilities.

You see, while AI may be incredibly powerful when it comes to processing vast amounts of data, identifying patterns, and optimizing for specific, narrowly defined goals, there are still many things that we humans are uniquely suited for - things like creativity, emotional intelligence, complex reasoning, and the ability to think beyond the immediate task at hand to consider broader ethical and societal implications.

And when we learn how to combine these human strengths with the raw processing power and efficiency of AI, magic can happen. We're talking about a level of problem-solving, innovation, and value creation that simply wouldn't be possible with either humans or machines working alone.

But of course, this is easier said than done. It's not like we can just snap our fingers and suddenly become expert collaborators with our AI counterparts. It's going to require a real mindset shift, as well as a commitment to continuously learning and upskilling ourselves in the areas that are most resistant to automation.

That's why, in this chapter, we're going to take a deep dive into what this human-AI symbiosis might look like in practice. We'll explore some of the key human traits and abilities that are likely to remain in high demand even as AI takes over more and more tasks - things like creativity, emotional intelligence, persuasion, and the ability to navigate complex social and cultural dynamics.

We'll also look at some real-world examples of humans and AI working together in powerful ways, from creative professionals using AI tools to augment their artistic abilities to business leaders leveraging AI insights to make better strategic decisions, to healthcare providers using AI to help diagnose and treat patients more effectively.

But we won't stop there. We'll also provide a practical roadmap for how you can start developing these key "complementary" skills and traits in your own life and career. We'll explore strategies for lifelong learning, building multidisciplinary expertise, and cultivating a mindset of curiosity and adaptability in the face of constant change.

And finally, we'll take a look at some of the emerging models and frameworks for effective human-AI collaboration - things like "centaur" chess teams, where human players work alongside AI to achieve super-human levels of play, or "cobots" in manufacturing, where humans and robots work side-by-side to build products more efficiently than either could alone.

By the end of this chapter, my hope is that you'll have a much clearer sense of how you can position yourself to thrive in the age of AI - not by fighting against the machines, but by learning to work with them in a spirit of collaboration and partnership.

Because here's the thing - the future is coming, whether we like it or not. And while it may be tempting to bury our heads in the sand or try to resist the inexorable march of technological progress, the truth is that the only way forward is through adaptation, learning, and growth.

So let's dive in and explore how we can build a future where humans and machines work together in harmony, leveraging the best of both worlds to create a more prosperous, innovative, and fulfilling society for all.

4.1. The Human-AI Symbiosis

First, let's talk about this idea of human-AI symbiosis - the notion that rather than seeing AI as a threat or a competitor, we should be thinking about how we can work together with these technologies in a complementary way, leveraging the unique strengths and abilities of both humans and machines.

Now, I know this might sound a bit abstract at first, but stick with me - because once you start to wrap your head around it, it's actually a really powerful and exciting concept.

So here's the basic idea: AI is incredibly good at certain things - things like processing huge amounts of data very quickly, identifying patterns and correlations that might be hard for humans to spot, and optimizing for very specific, well-defined goals. If you've ever used a recommendation algorithm on a streaming platform or seen a self-driving car navigate through traffic, you know what I'm talking about.

But here's the thing - as powerful as AI is, it's not good at everything. In fact, there are a lot of things that we humans are still way better at than machines - things like coming up with creative and original ideas, understanding and navigating complex social and emotional dynamics, and making sense of ambiguous or contradictory information.

And when you think about it, this makes perfect sense. After all, the human brain is an incredibly complex and sophisticated piece of biological machinery - the result of millions of years of evolution and adaptation. We have the ability to think in abstract and analogical ways, to draw insights from seemingly unrelated domains, and to make intuitive leaps that can lead to groundbreaking innovations and discoveries.

So the key to thriving in the age of AI is not to try to compete with machines on their own turf - in other words, to try to be faster, more efficient, or more precise at the kinds of tasks that AI is already really good at. Instead, it's to focus on developing and leveraging the skills and abilities that are uniquely human - the things that machines can't do (at least not yet).

And when we do that - when we combine the raw processing power and efficiency of AI with the creativity, empathy, and contextual understanding of humans - amazing things can happen.

Take the field of medicine, for example. In recent years, there's been a lot of hype around the idea of using AI to diagnose diseases and recommend treatments - and for good reason. AI algorithms can analyze vast amounts of patient data, spot patterns and correlations that might be hard for human doctors to see, and even predict which treatments are most likely to be effective for a given patient.

But here's the thing - no matter how advanced these AI systems become, they'll never be able to replace the human touch that's so essential to good medical care. A machine might be able to analyze a patient's symptoms and recommend a course of treatment, but it can't sit down with that patient, listen to their concerns and fears, and provide the kind of emotional support and guidance that can make all the difference in their recovery.

That's where the idea of human-AI symbiosis comes in. Imagine a future where doctors and AI work together seamlessly - where AI handles the routine tasks of data analysis and pattern recognition, freeing up human doctors to focus on the things that only they can do, like building relationships with patients, providing personalized care, and support, and using their intuition and creativity to come up with innovative solutions to complex medical problems.

Or take the world of art and design. In recent years, we've seen the emergence of all sorts of AI-powered tools that can help artists and designers work faster and more efficiently - things like algorithms that can generate color palettes or suggest design layouts based on a set of input parameters.

But as powerful as these tools are, they'll never be able to replace the human spark of creativity that lies at the heart of great art and design. Sure, an AI might be able to generate a perfectly optimized logo or website layout - but it can't come up with the kind of original, unexpected ideas that can really capture people's imaginations and emotions.

So again, the key is to think about how we can use AI as a tool to augment and enhance human creativity, rather than trying to replace it altogether. Imagine a world where designers and artists work alongside AI, using it to handle the more routine and technical aspects of their work, while they focus on the things that only human beings can do - things like coming up with wild, imaginative concepts, telling compelling stories, and connecting with audiences on a deep, emotional level.

Of course, this idea of human-AI symbiosis is not new. In fact, it's been around for decades - ever since the early days of computer science, when pioneers like J.C.R. Licklider and Douglas Engelbart first started thinking about how humans and machines could work together in powerful ways.

Licklider, in particular, was a big proponent of what he called "man-computer symbiosis" - the idea that by combining the strengths of human intelligence with the speed and precision of computers, we could achieve things that neither could do alone. He even wrote a famous paper on the topic back in 1960, where he envisioned a future where humans and machines would work together seamlessly, with computers handling the routine tasks of information processing and retrieval, while humans focused on the more creative and intuitive aspects of problem-solving.

And in many ways, that's exactly the future we're starting to see take shape today. From creative professionals using AI tools to augment their artistic abilities, to business leaders leveraging AI insights to make better strategic decisions, to healthcare providers using AI to help diagnose and treat patients more effectively - we're already starting to see the power of human-AI symbiosis in action.

But of course, there's still a long way to go. To really harness the full potential of this approach, we need to be intentional about developing the skills and mindsets that will allow us to work effectively alongside AI - things like creativity, emotional intelligence, adaptability, and the ability to think critically and ask the right questions.

And that's exactly what we'll be exploring in the rest of this chapter - how we can cultivate these key "human" skills and traits, how we can build effective models for human-AI collaboration, and how we can create a future where humans and machines work together in harmony to solve the world's biggest challenges. So let's keep diving in!

4.2. Nurturing Human Traits

Alright, so now that we've covered the idea of human-AI symbiosis and why it's so important, let's dive a little deeper into what that actually looks like in practice. Specifically, I want to focus on the key human traits and abilities that are likely to be most valuable and resistant to automation in the long run - what we might call our "complementary traits" or "complementary skills."

So what exactly do I mean by that? Well, think about it this way: as AI continues to advance and automate more and more tasks across more and more industries, there are going to be certain things that machines are really, really good at - things like processing huge amounts of data, identifying patterns, and optimizing for specific goals.

But there are also going to be certain things that machines struggle with - things that require a more nuanced, contextual, and intuitive understanding of the world. And it's in those areas where we humans have a real opportunity to shine - to leverage our unique strengths and abilities in ways that complement and enhance what machines can do.

So what are some of those key human complementary skills? Well, one of the big ones is emotional intelligence - the ability to perceive, understand, and manage emotions in ourselves and others. This is something that AI systems are still really bad at - even the most advanced sentiment analysis algorithms can only scratch the surface of the complex emotional dynamics that play out in human interactions.

But for us humans, emotional intelligence is a core part of who we are - it's what allows us to build strong relationships, negotiate complex social situations, and lead and inspire others. And in a world where more and more tasks are being automated, those skills are only going to become more valuable - because they're the things that machines simply can't replicate (at least not yet).

Another key human complementary skill is creativity - the ability to come up with original, unexpected ideas and solutions to problems. Again, this is something that machines struggle with - even the most advanced AI systems today are still essentially pattern-matching engines, recombining existing ideas in novel ways but not truly creating something new from scratch.

But for us humans, creativity is what drives innovation and progress - it's what allows us to imagine new possibilities, to see the world in fresh ways, and to come up with the kinds of wild, out-there ideas that can change everything. And in a world where more and more routine tasks are being automated, that kind of creative thinking is going to be more important than ever.

Other key human complementary skills include things like persuasion (the ability to influence and motivate others through language and storytelling), socio-cultural interpretation (the ability to navigate complex social and cultural dynamics), and moral reasoning (the ability to grapple with ethical dilemmas and make principled decisions in ambiguous situations).

All of these skills are deeply rooted in our human experience - they're the things that allow us to make sense of the messy, complicated world we live in, and to navigate it in ways that are meaningful and fulfilling. And while machines may be able to approximate some of these abilities to a certain degree, they'll never be able to fully replicate the depth and nuance of human understanding.

So what does all this mean for us as individuals? How can we make sure that we're developing and nurturing these key human complementary skills, so that we can thrive in a world where more and more tasks are being automated?

Well, one key thing is simply to be aware of what those skills are, and to start thinking about how we can cultivate them in ourselves. That might mean seeking out opportunities to build our emotional intelligence, whether through leadership roles, volunteering, or simply practicing empathy and active listening in our daily lives. It might mean carving out time for creative pursuits, whether that's writing, painting, music, or any other outlet that allows us to flex our imaginative muscles.

Another key thing is to start thinking about our education and career paths in a more holistic, multidisciplinary way. In a world where narrow technical skills are increasingly being automated, it's the people who can bring a diversity of perspectives and experiences to the table who are going to be most valuable. That might mean pursuing a liberal arts education that exposes us to a wide range of ideas and ways of thinking or seeking out opportunities to work across different fields and industries.

And of course, it also means staying curious and adaptable, and being willing to constantly learn and grow throughout our lives. Because the truth is, the skills that are valuable today may not be the same ones that are valuable tomorrow - and the people who are most successful in the long run will be the ones who are able to continually reinvent themselves and stay ahead of the curve.

Ultimately, the key thing to remember is that as powerful as machines are becoming, they're never going to be able to fully replicate the incredible complexity and richness of the human experience. We have a unique set of strengths and abilities that are deeply rooted in who we are as a species - and by nurturing and developing those abilities, we can put ourselves in the best possible position to thrive in a world where more and more tasks are being automated.

And the good news is, this isn't some far-off, science-fiction vision of the future - it's something that we can start working on right now, in our own lives and careers. By being intentional about developing our emotional intelligence, creativity, persuasion skills, and other key human complementary traits, we can position ourselves to be the kinds of leaders, innovators, and change makers that the world needs in the age of AI.

So what do you say? Are you ready to start cultivating your own unique set of human superpowers? Let's get to work!

4.3. Upskilling For The AI Age

Now that we've identified some of the key human super skills that are going to be most valuable in the age of AI, the next question is: how do we actually go about developing and cultivating those skills in ourselves?

It's a great question, and one that I think is going to be increasingly important for all of us to grapple with as we navigate this rapidly changing world of work. Because let's face it - the days of being able to coast through a 40-year career with the same narrow set of technical skills are pretty much over. If we want to stay relevant and competitive in the AI age, we're going to need to be constantly learning, growing, and adapting.

So what does that actually look like in practice? Well, I think there are a few key strategies that we can all start implementing in our own lives and careers.

First and foremost, I think we need to embrace the idea of life-long learning. And I don't just mean in the sense of taking a class here or there, or attending the occasional workshop. I mean really making learning a core part of our daily lives - setting aside dedicated time each day or each week to read, explore new ideas, and push ourselves outside our intellectual comfort zones.

One way to do that is to seek out opportunities for multidisciplinary learning - to expose ourselves to ideas and perspectives from fields outside our own areas of expertise. Because the truth is, some of the most innovative and groundbreaking ideas often come from the intersections between different disciplines - from applying concepts from one field to problems in another.

So if you're an engineer, maybe that means taking a class in psychology or design. If you're a marketer, maybe it means learning about data science or anthropology. The key is to be curious, and to always be looking for ways to expand your intellectual horizons.

Another key strategy is to focus on developing what I like to call "human-centric skills" - skills that are all about understanding and empathizing with other people. This could include things like active listening, emotional intelligence, storytelling, and the ability to build and maintain strong relationships.

Because here's the thing - no matter how advanced AI gets, there are always going to be certain things that machines simply can't do as well as humans. And a lot of those things have to do with our ability to connect with other people on a deep, emotional level - to understand their needs, their desires, their fears, and their dreams.

So if you want to future-proof your career, one of the best things you can do is to focus on developing those human-centric skills. That might mean taking an improv class to work on your ability to think on your feet and respond in the moment. It might mean studying psychology or neuroscience to better understand how people think and behave. Or it might mean volunteering for a nonprofit or community organization to build your empathy and your ability to work with diverse groups of people.

Of course, none of this is to say that technical skills aren't important - they absolutely are, and will continue to be. But I think the most successful people in the AI age will be those who are able to combine deep technical expertise with equally deep human skills.

Take the field of healthcare, for example. As AI becomes more sophisticated, we're likely to see more and more diagnostic and treatment tasks being automated. But that doesn't mean that doctors and nurses are going to become obsolete - far from it. If anything, their human skills will become even more important - their ability to communicate with patients, to provide emotional support and guidance, to make complex ethical decisions in the face of uncertainty.

And that's just one example - the same could be said for teachers, managers, salespeople, and really any profession that involves working with and understanding other people.

So if you're an undergraduate just starting to think about your career path, my advice would be this: don't just focus on developing a narrow set of technical skills. Instead, think about how you can cultivate a diverse range of human-centric skills that will allow you to thrive in a world where more and more tasks are being automated.

That might mean seeking out internships or extracurricular activities that allow you to work with people from different backgrounds and perspectives. It might mean taking classes in subjects like psychology, anthropology, or design thinking. Or it might mean starting your own project or initiative that allows you to flex your creative and entrepreneurial muscles.

The key is to be proactive, and to start thinking about your education and your career as a lifelong journey of growth and discovery. Because the truth is, none of us can predict exactly what the world will look like 10 or 20, or 50 years from now. But what we can do is stay curious, stay adaptable, and keep pushing ourselves to develop the kinds of timeless human skills that will always be in demand, no matter how much the world around us changes.

4.4. Soft Skills

In this chapter I'd like to focus on the secret weapon that's going to set you apart in the age of AI and automation: your incredible, irreplaceable, absolutely essential soft skills. Now, I know what you might be thinking - "soft skills? That sounds like something my grandma would put on her resume." But trust me, these babies are anything but fluffy.

First off, let's define what we mean by soft skills. Essentially, they're all the things that make you an awesome human being to work with - your communication chops, your ability to collaborate and play well with others, your knack for thinking creatively and solving problems in unique ways. They're the skills that help you build relationships, understand different perspectives, and navigate the complex social and emotional landscape of any workplace.

And here's the thing - as AI and automation take over more and more of the repetitive, routine tasks in our jobs, it's precisely these human-centric soft skills that are going to become more valuable than ever. After all, machines might be able to crunch numbers and analyze data like nobody's business, but they're still pretty terrible at things like empathy, persuasion, and thinking outside the box.

So what are some examples of the soft skills that are going to be your secret sauce in the AI age? Let's start with communication. And no, I'm not just talking about being able to string together a coherent email (although that's certainly important). I'm talking about the ability to listen actively, to ask great questions, to express yourself clearly and concisely, and to adapt your communication style to different audiences and situations.

Think about it - in a world where more and more of our interactions are happening via screens and algorithms, the ability to forge genuine human connections and get your point across in a way that resonates is going to be a serious superpower. Whether you're pitching a new idea to your boss, collaborating with a team of diverse personalities, or trying to persuade a room full of skeptical stakeholders, your communication skills are going to be the key to unlocking success.

But communication is just the tip of the iceberg when it comes to soft skills. Another big one is adaptability - the ability to roll with the punches, think on your feet, and pivot quickly in the face of change or uncertainty. And let's be real, if there's one thing we can count on in the age of AI, it's that change is going to be the only constant.

I mean, think about how quickly the world has transformed in just the past few decades. Back in the 90s, if you wanted to buy something, you had to actually leave your house and go to a store. If you wanted to learn a new skill, you had to sign up for a class or buy a book. And if you wanted to connect with someone on the other side of the world, you had to write them a letter and wait weeks for a response.

Fast forward to today, and we've got e-commerce, online learning, and instant global communication at our fingertips. And that's just the tip of the iceberg when it comes to the ways that technology is transforming our lives and our work. So if you want to thrive in this rapidly changing landscape, you've got to be able to adapt and evolve right along with it.

But perhaps the most important soft skill of all is something that might surprise you - empathy. That's right, the ability to put yourself in someone else's shoes, to understand their perspective and their needs, and to connect with them on a human level. And in a world where more and more of our interactions are being mediated by screens and algorithms, empathy is going to be the key to building the kind of deep, meaningful relationships that are the foundation of any successful career or endeavor.

Think about it - whether you're trying to collaborate with a team, sell a product or service, or even just navigate the complex social dynamics of your workplace, being able to understand and relate to the people around you is going to be absolutely essential. And as machines get better and better at handling purely logical and analytical tasks, it's the uniquely human ability to connect and empathize that's going to set you apart.

So how can you start cultivating these all-important soft skills? Well, the good news is that you don't need to go back to school or sign up for some expensive training program. A lot of it comes down to good old-fashioned practice and self-reflection.

Start by paying attention to the way you communicate and interact with others. Are you really listening to what they're saying, or are you just waiting for your turn to talk? Are you adapting your communication style to different situations and audiences, or are you a one-trick pony? And are you taking the time to put yourself in other people's shoes and see things from their perspective?

Another great way to build your soft skills is to seek out opportunities to collaborate and work with others. Whether it's joining a club or organization, volunteering for a cause you care about, or simply taking on more team projects at work or school, the more you practice working with diverse groups of people, the better you'll get at all those essential human skills.

And finally, don't be afraid to ask for feedback and guidance from others. Whether it's a trusted mentor, a colleague, or even just a friend or family member, getting an outside perspective on your strengths and areas for improvement can be incredibly valuable in helping you grow and develop as a person and a professional.

So there you have it - the secret weapon of soft skills. They might not be as flashy or buzzwordy as some of the other skills we've talked about in this book, but trust me, they're going to be absolutely essential in helping you navigate the brave new world of AI and automation. So start flexing those empathy muscles, brush up on your communication chops, and get ready to show the world just how valuable a little human touch can be.

4.5. Human-AI Collaboration Models

We've talked a lot about the importance of developing human super skills to thrive in the age of AI. But let's be honest - no matter how emotionally intelligent or creative, or adaptable we become, there's no denying that machines are going to be taking on more and more tasks that were once the exclusive domain of humans.

So the question becomes: how can we make sure that this transition is a collaborative one, rather than an adversarial one? How can we position ourselves to work alongside AI in a way that maximizes both human and machine capabilities, rather than pitting them against each other?

Well, that's exactly what we're going to explore in this final sub-chapter. We're going to look at some tangible models and frameworks for effective human-AI collaboration - ways of structuring our work and our decision-making processes to harness the power of both human and machine intelligence.

One key concept that I think is really useful here is the idea of humans as "stratifiers" and AI as "satisfiers." Let me break that down a bit.

Essentially, the idea is that humans are really good at setting high-level objectives and priorities - at looking at the big picture and determining what outcomes we want to achieve. We're the ones who can say, "Okay, our goal is to increase customer satisfaction by 10% this quarter," or "We want to reduce our carbon footprint by 50% over the next five years."

But once we've set those high-level objectives, AI can be incredibly powerful at helping us figure out how to actually achieve them. It can crunch through massive amounts of data, identify patterns and insights that we might miss, and optimize processes and systems to help us reach our goals more efficiently and effectively.

So in this model, humans are the stratifiers - the ones setting the overall direction and priorities. And AI is the satisfier - the tool that helps us execute on those priorities in the most optimized way possible.

But of course, it's not quite as simple as just setting a goal and letting the machines take over. There needs to be a constant feedback loop between humans and AI - a way for us to monitor progress, adjust course when needed, and make sure that the AI is actually aligned with our objectives.

One way to do that is through something called a responsibility assignment matrix, or RACI matrix for short. Essentially, this is a tool that helps clarify roles and responsibilities in any given project or decision-making process.

In a human-AI context, it might look something like this:

- Humans are "responsible" for setting the overall objectives and making final decisions.

- AI is "accountable" for executing those objectives and providing data-driven insights and recommendations.

- Humans are "consulted" throughout the process to provide context, domain expertise, and ethical oversight.

- Both humans and AI are "informed" of progress and results at regular intervals.

By clearly defining these roles and responsibilities upfront, we can create a more seamless and effective collaboration between humans and machines.

Another key aspect of effective human-AI collaboration is what I like to call "augmented decision-making." This is the idea that rather than just relying on either human judgment or AI algorithms in isolation, we can combine the two in powerful ways to make better decisions overall.

For example, let's say you're a doctor trying to diagnose a rare disease. You might start by reviewing the patient's symptoms and medical history, drawing on your own expertise and intuition. But you could also feed that information into an AI system that has been trained on millions of similar cases, and get a set of probabilistic recommendations back.

Then, rather than just blindly following the AI's advice, you could use it as a starting point for further investigation - a way to surface possibilities you might not have considered or to validate your own hypotheses. And ultimately, you as the human expert would make the final call, taking into account both the AI's insights and your own professional judgment.

This kind of augmented decision-making process has the potential to be incredibly powerful - not just in healthcare, but in fields like finance, marketing, product development, and beyond. By leveraging the strengths of both human and machine intelligence, we can make decisions that are more informed, more accurate, and more impacting than either could achieve alone.

Of course, none of this is to say that the transition to greater human-AI collaboration will be a smooth or easy one. There are certainly risks and challenges involved - from issues of bias and fairness in AI algorithms, to concerns about job displacement and economic disruption.

But I firmly believe that if we approach this transition thoughtfully and proactively - if we focus on developing our own human super skills, while also putting in place the right frameworks and processes for effective collaboration - then we have the potential to create a future in which humans and machines work together in truly symbiotic ways.

In fact, we can look to history for inspiration here. Think about the Industrial Revolution of the 18th and 19th centuries, when new technologies like the steam engine and the cotton gin transformed the nature of work and society as a whole.

At first, these technologies were met with fear and resistance - there were concerns about job losses, about the de-skilling of labor, and about the erosion of traditional ways of life. But over time, as societies adapted and evolved, these technologies ended up being powerful engines of growth and prosperity - they created whole new industries and job categories, and raised living standards for millions of people around the world.

I believe that we're on the cusp of a similar transformation today with the rise of AI. And just like in the Industrial Revolution, the key to successfully navigating this transformation will be a willingness to adapt, to learn, and to collaborate in new ways.

Together let's embrace this moment of change and possibility. Let's focus on developing our own human super skills, while also building the frameworks and the mindsets needed for effective human-AI collaboration. And let's work together to create a future in which humans and machines can truly thrive side-by-side.

CHAPTER 5

CULTIVATING
AI-RESISTANT SKILLS

As we dive deeper into this journey of navigating the AI revolution, it's time to focus on what truly sets us apart as humans. In a world where machines are getting smarter by the day, how can we ensure that we're not just keeping up, but thriving? The answer lies in cultivating and mastering skills that are resistant to automation - abilities that showcase the unparalleled power of the human mind and spirit.

In this chapter, we're going to explore four key areas that highlight our unique human capabilities: creativity, emotional intelligence, complex communication, and ethical reasoning. These are the superpowers that will help you stand out in an AI-driven world and make you an indispensable asset in any field you choose to pursue.

First up, we have creativity - the ultimate human skill. Sure, AI can generate art, music, and even write stories, but can it truly innovate like a human can? We'll dive into the cognitive processes behind creative thinking, like analogical reasoning and conceptual exploration, and explore how our emotions and experiences fuel our ability to come up with original, groundbreaking ideas. And don't worry, we'll also provide practical tips on how to boost your creative capabilities, from idea generation techniques to nurturing your curiosity.

Next, we'll tackle emotional intelligence, which is basically AI's Achilles heel. While machines can mimic human emotions, they still struggle with genuinely understanding and responding to them appropriately. We'll explore the neuroscience behind empathy and other social cognition abilities that are deeply rooted in our biology as humans. And we'll provide guidance on how to improve your emotional quotient (EQ) through practices like self-reflection and developing your "person-skills".

But being able to come up with brilliant ideas and connect with others emotionally is only half the battle - you also need to be able to communicate those ideas effectively. That's where complex communication mastery comes in. We'll delve into strategies for enhancing your ability to convey context-rich ideas across multiple domains, from storytelling and metaphorical thinking to interpreting subtext and tailoring your message for different audiences. These are skills that will set you apart as a leader and influencer, no matter what career path you choose.

Finally, we'll explore one of the most quintessentially human domains: ethical and moral reasoning. As machines become more integrated into our lives and decision-making processes, it's critical that we have individuals who can navigate the complex ethical landscape that comes with it. We'll discuss why skills like perspective-taking, dealing with ambiguities, and reasoning about broader consequences are so important, and provide guidance on how to develop your own moral compass through practices like studying philosophy and cultivating civic literacy.

Now, I know what you might be thinking - this all sounds great, but how do I actually develop these skills? Won't it take years of practice and experience? Well, yes and no. While it's true that mastering these abilities takes time and effort, the good news is that you can start building them right now, no matter where you are in your academic or professional journey.

Whether it's taking a class on creative writing, volunteering for a cause you care about, or simply engaging in more meaningful conversations with people from different backgrounds, there are countless ways to start cultivating your AI-resistant skills today. And the best part? These are the kinds of skills that will serve you well no matter what the future holds - whether you end up working alongside AI or not.

So get ready to unleash your human potential and show the world what you're truly capable of. The AI revolution may be coming, but with the right skills and mindset, you'll be more than ready to meet it head-on. Let's start building those superpowers together!

5.1. Creativity: The Ultimate Human Skill

Alright, let's dive into the wonderful world of creativity - the ultimate human skill that sets us apart from even the most advanced AI systems out there. Now, I know what you might be thinking: "But wait, I've seen AI create art, music, and even write stories. How can creativity be uniquely human?" Well, my friend, let me tell you - there's a big difference between mimicking creative output and truly embodying the cognitive processes that drive original, innovative thinking.

At its core, creativity is all about coming up with novel and useful ideas - ones that haven't been thought of before, and that have the potential to change the game in whatever field you're working in. It's about connecting dots that others haven't connected and seeing the world in a way that's entirely your own. And while AI might be able to generate content that looks or sounds creative on the surface, it's still ultimately based on patterns and data from existing works - it's not truly creating something from scratch.

So what is it about the human mind that makes us so uniquely suited for creative thinking? Well, it turns out there are a few key cognitive processes that come into play. One of the big ones is analogical reasoning - the ability to draw connections between seemingly unrelated concepts or ideas. This is what allows us to take knowledge or experiences from one domain and apply them to another in a novel way - like how the inventor of Velcro was inspired by the way burrs stuck to his dog's fur during a hike.

Another key aspect of creativity is conceptual exploration - the ability to mentally manipulate and play with ideas, testing out different combinations and scenarios to see what emerges. This is where our imagination really comes into play - we can envision things that don't exist yet, and explore the possibilities of what could be. And often, it's in those moments of unstructured play and exploration that the most innovative ideas are born.

But perhaps the most important factor in human creativity is the role of emotions and experiences. Our unique perspectives, shaped by our individual histories, cultures, and ways of seeing the world, are what allow us to bring something truly original to the table. When we create, we're not just drawing on intellectual knowledge - we're tapping into our deepest feelings, desires, and fears, and using them as fuel for our ideas.

Take the example of Frida Kahlo, the iconic Mexican painter whose work was deeply influenced by her own life experiences - from the physical and emotional pain she endured after a bus accident, to her tumultuous relationship with fellow artist Diego Rivera. Kahlo's paintings are famous for their raw, honest portrayal of the female experience, and for the ways in which they challenge traditional notions of beauty and femininity. It's hard to imagine an AI system being able to create something so deeply personal and resonant because it simply doesn't have the same lived experiences to draw from.

So how can we cultivate our own creative capabilities, and make sure we're flexing those uniquely human muscles? One key strategy is to engage in regular idea generation techniques - things like brainstorming, mind-mapping, and free-writing, which allow us to explore new concepts and connections in a non-judgmental way. The key is to let your mind wander and see where it takes you - you never know what kind of brilliant idea might emerge when you give yourself permission to think outside the box.

Another important factor is nurturing your curiosity - being open to new experiences, perspectives, and ways of seeing the world. The more diverse your range of interests and knowledge, the more material you have to draw from when it comes to creative thinking. So don't be afraid to explore topics or hobbies that might seem unrelated to your main area of focus - you never know how they might inform your work in unexpected ways.

And finally, don't underestimate the power of engaging in creative hobbies outside of your professional life. Whether it's painting, writing, music, or any other form of artistic expression, having a creative outlet can help keep those imaginative muscles in shape and provide a space for you to explore new ideas without the pressure of external expectations.

Of course, this is just scratching the surface of what makes human creativity so special and irreplaceable. Throughout history, we've seen countless examples of individuals whose unique creative visions have transformed the world - from Leonardo da Vinci's groundbreaking inventions and artwork to Marie Curie's pioneering discoveries in radio-activity, to Steve Jobs' revolutionary designs for Apple products.

What all of these creators had in common was a willingness to think differently, to challenge assumptions, and to bring their whole selves - emotions, experiences, and all - to their work. And in a world where AI is increasingly able to automate many tasks and processes, it's those deeply human qualities that will continue to set us apart and drive innovation forward.

So embrace your creativity, and don't be afraid to let it shine. The world needs your unique perspective and ideas now more than ever - and with the right tools and mindset, there's no limit to what you can create.

5.2. Emotional Intelligence: AI's Achilles Heel

Now let's move on to talk about one of the most fascinating and complex aspects of the human experience - our emotions, and the power they hold in shaping our interactions with each other and the world around us. It's no secret that we're deeply emotional creatures - from the joy we feel when we accomplish a long-held goal, to the sadness that comes with losing a loved one, to the fear and anger that can arise in the face of injustice or threat. Our emotions are what make us human, and they're also what make us incredibly difficult to replicate in artificial form.

Sure, AI systems can be programmed to mimic certain emotional displays - think of virtual assistants like Siri or Alexa, who can respond with pre-written phrases that sound empathetic or understanding. But when it comes to truly understanding and responding to the nuances of human emotion in real time, AI still has a long way to go. That's because emotional intelligence - the ability to be aware of, manage, and respond appropriately to both our own emotions and those of others - is an incredibly complex skill that's deeply rooted in our biology and evolutionary history.

At a basic level, our emotions are controlled by a set of brain structures known as the limbic system, which includes regions like the amygdala, hippocampus, and hypothalamus. These areas are responsible for processing sensory information, forming memories, and generating the physiological responses that we associate with different emotional states - things like changes in heart rate, blood pressure, and hormone levels. But our emotions are also heavily influenced by our past experiences, our beliefs and values, and our social and cultural contexts - all of which shape the way we interpret and respond to different situations.

One of the key components of emotional intelligence is empathy - the ability to understand and share the feelings of another person. And this is where AI really struggles to keep up with us humans. Because empathy isn't just about recognizing and labeling different emotional states - it's about being able to put yourself in someone else's shoes, to imagine what they might be thinking or feeling, and to respond in a way that shows genuine understanding and care.

Think about a time when you've gone to a friend or loved one for support during a difficult moment - maybe you were feeling overwhelmed with stress or dealing with a painful breakup or loss. What made that interaction feel genuinely supportive and comforting? Chances are, it wasn't just the words that were said, but how they were delivered - the tone of voice, the body language, the sense that the other person was truly listening and trying to understand your perspective.

That kind of deep, empathetic connection is something that humans are uniquely suited for, thanks to our long evolutionary history as social creatures. We've developed a complex set of neural networks and cognitive abilities that allow us to read subtle cues in facial expressions, voice inflections, and body language, and to respond in ways that show care and concern for others. And while AI systems might be able to analyze those cues on a surface level, they don't have the same deep well of personal experience and context to draw from in crafting a truly empathetic response.

So what can we do to cultivate and strengthen our own emotional intelligence skills, and make sure we're not relying too heavily on AI to navigate the complexities of human interaction? One key strategy is to practice regular self-reflection - taking time to check in with yourself, to notice and name your own emotions, and to explore the thoughts and beliefs that might be driving them. This kind of self-awareness is crucial for being able to manage and regulate your own emotions in healthy ways, and for developing greater empathy and understanding for others.

Another important tool is mindfulness - the practice of being present and fully engaged in the current moment, without judgment or distraction. By learning to tune into your own bodily sensations, thoughts, and feelings in a non-judgmental way, you can develop greater clarity and calm in the face of difficult emotions, and be more responsive and adaptable in your interactions with others.

And of course, there's no substitute for good old-fashioned "people skills" - the ability to listen actively, to communicate clearly and effectively, and to build strong, trusting relationships with others. Whether it's through practicing active listening in conversations, seeking out feedback from others, or engaging in social activities and volunteer work, there are countless ways to strengthen your interpersonal muscles and become more attuned to the needs and perspectives of those around you.

The good news is that emotional intelligence is a skill that can be developed and strengthened over time, just like any other ability. And in a world where more and more of our interactions and decisions are being mediated by technology, it's becoming increasingly important to make sure we're not losing touch with our own humanity in the process.

After all, just think about some of the most transformative leaders and changemakers throughout history - people like Mahatma Gandhi, Martin Luther King Jr., or Mother Teresa. What set them apart wasn't just their intellectual brilliance or strategic savvy, but their deep capacity for empathy, compassion, and understanding. They were able to connect with people on a profound level, to inspire and mobilize them towards a common cause, and to create lasting change in the world.

And that's the kind of emotional intelligence that we should all be striving for - not just in our personal lives, but in our work and leadership as well. Because in a world that's becoming increasingly automated and digitized, it's the uniquely human skills of empathy, creativity, and connection that will ultimately set us apart and help us thrive.

5.3. Complex Communication Mastery

Next let's discuss complex communication - the art of expressing ideas, emotions, and perspectives in a way that's clear, compelling, and deeply human. Now, I know what you might be thinking - in a world where AI can write news articles, compose poetry, and even generate entire conversations, what's so special about human communication? Can't we just let the machines do the talking for us?

Well, not so fast! While it's true that AI has made some impressive strides in language processing and generation, there's still a wide gulf between the kind of communication that machines can produce and the rich, nuanced, and context-dependent communication that we humans engage in every day.

Think about it - when you're having a conversation with a friend, a colleague, or a loved one, you're not just exchanging information in a vacuum. You're drawing on a complex web of shared experiences, cultural references, and emotional cues to convey your thoughts and feelings in a way that resonates with your audience. You're using tone, body language, and other nonverbal signals to add depth and meaning to your words. And you're constantly adapting your message based on the reactions and feedback you're getting from the other person, in real-time.

That kind of dynamic, contextual communication is something that even the most advanced AI systems struggle with. Sure, they can generate grammatically correct sentences and even mimic certain styles or tones based on patterns in their training data. But they don't have the same deep understanding of the human experience that allows us to communicate with true empathy, nuance, and persuasive power.

Take the example of a great public speaker - someone like Martin Luther King Jr., whose "I Have a Dream" speech remains one of the most powerful and influential pieces of oratory in American history. What made that speech so effective wasn't just the words themselves, but the way in which they were delivered - the passion and conviction in King's voice, the vivid imagery and metaphorical language he used to paint a picture of a better future, the way he wove together themes of justice, equality, and human dignity to create a compelling narrative that spoke to the hearts and minds of his audience.

That kind of masterful communication is the result of a complex interplay of skills and abilities that are uniquely human. It requires a deep understanding of the audience and the context in which the message is being delivered, as well as the ability to craft a narrative that resonates on both an intellectual and emotional level. It involves using metaphorical and analogical thinking to make abstract ideas more concrete and relatable, and to draw connections between seemingly disparate concepts. And it often involves reading between the lines and interpreting the subtext of a situation - picking up on subtle cues and unspoken assumptions that can be just as important as the words themselves.

So how can we cultivate these complex communication skills in ourselves, and make sure we're not losing touch with the power of human expression in an age of AI-generated content? One key strategy is to focus on developing our storytelling abilities - learning how to craft compelling narratives that engage and inspire others. This might involve studying great works of literature, film, or theater to see how master storytellers use language, structure, and imagery to create emotional resonance and convey deeper truths about the human experience.

Another important skill is metaphorical thinking - the ability to use analogies, similes, and other figurative language to make complex ideas more accessible and memorable. By drawing connections between abstract concepts and concrete, sensory experiences, we can help others grasp the essence of an idea in a way that sticks with them long after the conversation is over.

And of course, there's no substitute for good old-fashioned active listening and empathy - the ability to put yourself in someone else's shoes, to understand their perspective and experiences, and to tailor your message in a way that speaks directly to their needs and concerns. This might involve practicing mindfulness and self-reflection, cultivating curiosity about others' lives and backgrounds, and seeking out opportunities to engage with people from different cultures and walks of life.

The good news is that these complex communication skills are ones that we can all develop and strengthen over time, through practice, feedback, and a willingness to step outside our comfort zones. And in a world where more and more of our interactions are mediated by screens and algorithms, they're becoming increasingly valuable and important.

After all, just think about some of the most influential communicators throughout history - people like Winston Churchill, whose stirring speeches helped rally the British people during World War II; or Maya Angelou, whose poetry and memoirs gave voice to the experiences of black women in America; or even modern-day figures like Oprah Winfrey or Brené Brown, who have used their platforms to inspire millions with messages of empowerment, vulnerability, and self-acceptance.

What sets these communicators apart isn't just their mastery of language or their ability to generate clever turns of phrase - it's their deep understanding of the human experience, their empathy and emotional intelligence, and their willingness to speak truth to power in a way that moves and transforms us.

And that's the kind of complex communication that we should all be striving for - not just in our personal lives, but in our work and leadership as well. Because in a world that's becoming increasingly automated and impersonal, it's the uniquely human capacity for connection, understanding, and inspiration that will ultimately set us apart and help us thrive.

5.4. Effective Learning Methods

It's time to talk about something that's near and dear to my heart - effective learning methods! Because the reality is, in a world that's changing as fast as ours, the ability to learn and adapt is pretty much the most important skill you can have. And if you want to stay ahead of the curve and thrive in the age of AI, you're going to need to be intentional and strategic about how you approach your own learning and development.

You might think something like - "Ugh, learning. Isn't that just about sitting in a boring classroom, memorizing a bunch of facts and formulas?" Well, I'm here to tell you that it doesn't have to be that way. In fact, some of the most powerful and effective learning methods out there are all about getting hands-on, getting creative, and getting real.

Take project-based learning, for example. This is all about applying the skills and knowledge you're acquiring to real-world scenarios and challenges. Instead of just learning about something in the abstract, you're actually putting it into practice and seeing how it works (or doesn't work) in a practical context.

Imagine you're an engineering student learning about robotics. Instead of just reading about the principles of machine learning and computer vision in a textbook, what if you actually got to build and program your own robot as part of a class project? Not only would you be learning by doing, but you'd also be developing valuable skills like problem-solving, teamwork, and project management - skills that will serve you well no matter what field you end up in.

Or think about collaborative learning and group discussions. When you learn alongside others, you're not just absorbing information - you're actively engaging with it, questioning it, and building on it together. You're exposed to different perspectives and ideas, and you have the opportunity to learn from your peers as well as your instructors.

This is something that's been known for centuries - just look at the ancient Greek philosophers like Socrates and Plato, who believed that the best way to arrive at truth and knowledge was through dialogue and debate. They would gather in public spaces like the Agora and engage in lively discussions and arguments, challenging each other's ideas and assumptions in order to arrive at a deeper understanding.

And let's not forget about the power of reflective journaling and self-assessment. It's one thing to learn something in the moment, but it's another thing entirely to take the time to reflect on what you've learned, how it connects to your own experiences and goals, and what you still need to work on. By regularly checking in with yourself and assessing your own progress, you can stay motivated, stay on track, and make sure you're getting the most out of your learning journey.

This is something that's been proven time and again in research on learning and education. Studies have shown that students who engage in reflective practices like journaling and self-assessment tend to have better learning outcomes, higher levels of engagement and motivation, and even better mental health and well-being.

But of course, effective learning isn't just about what happens inside your own head - it's also about the resources and opportunities you have access to. And lucky for us, there are more ways than ever to acquire new skills and knowledge, thanks in large part to the wonders of technology.

Take online learning platforms like Coursera, edX, and Udemy - these are incredible resources that give you access to courses and programs from some of the top universities and experts in the world, all from the comfort of your own home (or dorm room, or coffee shop, or wherever you happen to be). Whether you're looking to pick up a new programming language, learn about the latest developments in AI and machine learning, or just explore a topic that interests you, there's an online course out there for you.

But online learning isn't the only way to acquire new skills and knowledge - there are also things like mentorship programs, networking opportunities, and workshops and seminars focused on specific topics like creativity, emotional intelligence, and complex communication. These are all great ways to learn from others who have been where you are, and to build relationships and connections that can help you grow both personally and professionally.

And let's not forget about the incredible potential of emerging technologies like virtual reality and augmented reality for learning and skill acquisition. Imagine being able to step inside a virtual laboratory and conduct experiments in a safe and controlled environment, or to practice public speaking in front of a simulated audience that feels just like the real thing. The possibilities are endless, and they're only going to keep expanding as the technology advances.

So there you have it - a whirlwind tour of some of the most effective learning methods out there, and the incredible resources and opportunities available to help you acquire the skills and knowledge you need to thrive in the age of AI. But of course, this is just the tip of the iceberg - there are so many other ways to learn and grow, and what works best for you might be different from what works for someone else.

The key is to be curious, to be proactive, and to never stop learning. Whether you're a student just starting out on your educational journey, or a seasoned professional looking to stay sharp and relevant in a rapidly changing world, the ability to learn and adapt is the ultimate competitive advantage. So go out there and start exploring - you never know what you might discover, or where it might take you. The future belongs to the lifelong learners - and with the right mindset and the right tools, that future will be yours.

5.5. Ethical And Moral Reasoning

Lastly, one of the most important and complex skills that we as humans possess is our ability to engage in ethical and moral reasoning. This is the process of figuring out what's right and wrong, what's just and unjust, and how we ought to behave in a world full of competing values, interests, and priorities. And it's a skill that's becoming increasingly important in an age where machines are taking on more and more decision-making responsibilities.

Now, I know what you might be thinking - can't we just program AI systems to follow a set of ethical rules and principles? Can't we just input a bunch of data on human values and societal norms and let the machines figure out the rest? Well, it turns out that it's not quite that simple.

You see, ethical reasoning is a deeply human skill that's rooted in our ability to take multiple perspectives, to grapple with ambiguity and contradiction, and to think deeply about the broader consequences of our actions. It's not just about following a set of pre-programmed rules or algorithms - it's about using our judgment, our empathy, and our sense of moral intuition to navigate complex situations and make hard choices.

Think about a classic ethical dilemma like the trolley problem - a thought experiment where you're asked to imagine a runaway trolley barreling down a track toward five people. You have the ability to pull a lever and divert the trolley to a different track, but there's one person standing on that track who would be killed as a result. What do you do? Do you take action and sacrifice one life to save five, or do you do nothing and let nature take its course?

There's no easy answer to this question, and different people might come to different conclusions based on their own moral values and principles. Some might argue that taking action is the right thing to do, as it minimizes overall harm and suffering. Others might say that actively causing someone's death is never justified, no matter the consequences.

And still others might grapple with questions of intention, responsibility, and the unintended consequences of our choices.

The point is, there's no simple algorithm or formula that can solve these kinds of ethical dilemmas for us. It requires us to use our human judgment, to weigh competing values and principles, and to make difficult trade-offs based on our own sense of what's right and wrong. And that's something that even the most advanced AI systems struggle with.

Sure, we can try to program ethical principles into machines - things like the three laws of robotics proposed by science fiction writer Isaac Asimov, which stipulate that robots must not harm humans, must obey human orders, and must protect their own existence. But as anyone who's read Asimov's stories knows, even these seemingly straightforward rules can lead to all sorts of unintended consequences and moral quandaries.

For example, what happens when two of the laws come into conflict with each other? What if obeying a human order would cause harm to another human? What if protecting a robot's own existence would mean failing to prevent greater harm? These are the kinds of complex ethical questions that require nuanced human judgment to navigate.

And it's not just in hypothetical scenarios that these issues arise. In the real world, we're already seeing the challenges of trying to encode human values and ethics into machines. Take the example of self-driving cars - a technology that has the potential to save countless lives by reducing traffic accidents, but that also raises thorny ethical questions about how to prioritize different lives in the event of an unavoidable collision.

Should the car be programmed to always prioritize the safety of its passengers, even if that means putting pedestrians or other drivers at risk? Should it make decisions based on the age, health status, or perceived social value of the people involved? These are the kinds of moral calculations that humans struggle with, and that machines are ill-equipped to handle on their own.

So what can we do to cultivate our own ethical reasoning skills, and ensure that we're equipped to grapple with these complex moral questions in an age of increasing automation? One key approach is to study philosophy and ethics - to engage with the long history of human thought on questions of right and wrong, justice and injustice, and the nature of the good life.

By reading the works of great moral philosophers like Aristotle, Kant, and Mill, we can develop a deeper understanding of different ethical frameworks and principles, and learn to apply them to real-world situations. We can also engage in moral case studies and thought experiments, like the trolley problem, to sharpen our ethical reasoning skills and explore the implications of different courses of action.

Another important approach is to cultivate a sense of civic literacy and engagement - to stay informed about the social and political issues of our time, and to actively participate in shaping the values and norms of our communities. This might involve volunteering for local organizations, attending town hall meetings or public forums, or simply engaging in respectful dialogue and debate with others who hold different views.

The key is to recognize that ethical reasoning is not something that can be outsourced to machines or algorithms - it's a fundamentally human skill that requires ongoing practice, reflection, and growth. And in a world where the decisions we make can have far-reaching consequences for ourselves and others, it's a skill that we can't afford to neglect.

So let's commit to being lifelong students of ethics and moral philosophy, to grappling with the big questions of right and wrong, and to using our uniquely human capacities for empathy, reason, and judgment to navigate the complex challenges of our time. Because in the end, it's not just about being good at our jobs or staying relevant in an age of automation - it's about being good humans and leaving the world a little bit better than we found it.

CHAPTER 6

PERSONAL BRANDING FOR AN AI WORLD

So far we've explored the incredible potential of artificial intelligence and how it's transforming industries, skills, and even the very nature of work itself. But as we've seen, with this great technological change comes great personal responsibility - the responsibility to adapt, upskill, and position ourselves to thrive in this brave new world.

And that's where the idea of personal branding comes in. Now, I know that might sound like some fancy marketing jargon or Instagram influencer stuff. But at its core, personal branding is really about understanding what makes you uniquely valuable as a human being, and then communicating that value effectively to the world around you.

In the age of AI, this is going to be more important than ever. Because let's face it - as machines get smarter and more capable, a lot of the skills and abilities that we've traditionally relied on to make a living are going to become increasingly automated. If your whole value proposition is based on being able to crunch numbers or follow a set of rules, then you might find yourself struggling to stay relevant in a world where algorithms can do that faster, cheaper, and more accurately than any human ever could.

But here's the good news - as we've explored throughout this book, there are certain things that humans can do that machines can't (at least not yet). Things like creativity, emotional intelligence, complex reasoning, and the ability to connect with others on a deep, personal level. These are the skills and abilities that are going to be in high demand in the AI era - the ones that will help you stand out from the crowd and make yourself indispensable in any field or industry.

And that's really what this chapter is all about - helping you identify and cultivate your own unique "human edge," and then build a personal brand around it that showcases your value to the world.

We'll start by guiding you through a process of self-reflection and assessment to identify your core strengths and abilities. This isn't about just listing out your technical skills or job experiences - it's about really digging deep and understanding what makes you tick as a person. What are the things that you're naturally gifted at, that energize you and bring you joy? What are the qualities and characteristics that others consistently praise you for? What are the experiences and perspectives that have shaped who you are and how you see the world?

From there, we'll explore how to take those core strengths and translate them into a compelling personal brand that's tailored for the AI age. This might involve crafting a powerful personal mission statement that articulates your unique value proposition or building an online presence that showcases your AI-complementary skills and experiences. We'll look at case studies of people who have successfully pivoted their personal brands to thrive in the age of automation and draw lessons and inspiration from their stories.

But of course, a personal brand is only as valuable as the relationships and opportunities it helps you build. So we'll also dive into the art and science of networking in the AI era - how to forge meaningful connections with others, both online and off, and leverage those relationships to open up new doors and possibilities. We'll explore the increasing importance of social and emotional skills in a world where transactional tasks are being automated away, and provide practical tips and strategies for building your own relationship capital.

Finally, we'll get tactical with some concrete tips and techniques for marketing your human skills and abilities in the AI age. From optimizing your resume and online profiles to showcase your unique value, to acing interviews and pitches by highlighting your AI-complementary strengths, we'll arm you with the tools and frameworks you need to put your best foot forward and make yourself impossible to ignore.

Now, I know that some of this stuff might feel a little overwhelming or even uncomfortable at first. Self-promotion and personal branding can sometimes get a bad rap, and it's easy to feel like you're somehow being inauthentic or "selling out" by focusing on marketing yourself.

But the truth is, in a world that's changing as rapidly and profoundly as ours is, being intentional and proactive about your personal brand isn't just a nice-to-have - it's an absolute necessity. If you want to thrive in the age of AI and automation, you need to be crystal clear on what makes you uniquely valuable as a human being, and then shout that from the rooftops (or at least from your LinkedIn profile).

And the best part is, by doing this work of self-discovery and personal branding, you're not just setting yourself up for career success - you're also gaining a deeper understanding of who you are, what you stand for, and what kind of impact you want to have on the world. And that's a powerful thing, no matter what the future holds.

So get ready to dig deep, get creative, and start building your own personal brand for the AI age. The world is waiting for you to show up and make your mark - and with the right tools and mindset, there's no limit to what you can achieve.

6.1. Discovering Your Human Edge

Alright, let's dive into the juicy stuff - figuring out what makes you, you! In a world where AI is getting smarter by the day, it's more important than ever to have a crystal-clear understanding of your own unique strengths and superpowers. Because honestly - if you're trying to compete with machines on skills like crunching numbers or memorizing facts, you're gonna have a bad time.

But here's the good news - there are so many things that we humans are uniquely gifted at, that even the most advanced AI systems can't touch (at least not yet). And by identifying and cultivating those skills and abilities, you can position yourself to thrive in the age of automation and beyond.

So how do you go about discovering your own personal "human edge"? Well, it starts with a little bit of self-reflection and introspection. I know, I know - that might sound like some woo-woo self-help stuff. But trust me, taking the time to really understand what makes you tick is one of the most valuable investments you can make in your future success.

One way to start is by thinking about the things that you're naturally drawn to and energized by. What are the activities or topics that you could talk about for hours without getting bored? What are the projects or roles that you've taken on in the past that just felt like a perfect fit, like you were exactly where you were meant to be?

For some people, it might be all about creativity and self-expression - things like writing, art, music, or design. For others, it might be more about connecting with people and building relationships - skills like empathy, communication, and emotional intelligence. And for others still, it might be about tackling complex problems and coming up with innovative solutions - things like critical thinking, analysis, and strategic reasoning.

The key is to not just focus on what you're good at in a narrow, technical sense, but to really think about the deeper qualities and characteristics that define who you are as a person. Are you someone who's always been naturally curious and loves learning new things? Someone who's a great listener and can make anyone feel heard and understood? Someone who's a natural leader and can rally people around a common cause?

These are the kinds of skills and abilities that are going to be increasingly valuable in an AI-driven world - because they're the things that machines can't easily replicate or automate away. And the more you can understand and articulate what makes you uniquely human and valuable, the better positioned you'll be to thrive in the years ahead.

Now, I know that some of you might be thinking - "But wait, I'm not some super-creative artistic genius or charismatic people person! What if I don't have any special human skills to offer?" And to that I say - nonsense! Every single one of us has unique strengths and abilities that are valuable and needed in the world. It's just a matter of taking the time to discover and cultivate them.

And that's where some good old-fashioned assessments and self-reflection exercises can come in handy. There are all kinds of tools and frameworks out there that can help you get a better understanding of your own strengths and skills - from personality tests like the Myers-Briggs or Enneagram to more career-focused assessments like Strengths-Finder or the MAPP (Motivational Appraisal of Personal Potential).

But you don't necessarily need some fancy assessment to start discovering your human edge. Sometimes, it's just about paying attention to your own experiences and the feedback you get from others. Think about the times in your life when you've felt most alive, most in flow, most like you were doing exactly what you were meant to be doing. What were the common threads or themes that ran through those experiences?

Or think about the compliments and praise that you consistently receive from others - the things that people say you're naturally gifted at or that you bring to the table in any situation. Sometimes, it's easy for us to brush off compliments or downplay our own strengths - but those external reflections can be incredibly valuable data points in understanding what makes us unique and valuable.

The goal here is not to come up with some perfectly polished, LinkedIn-ready list of skills and strengths (although that can certainly be a helpful exercise too). It's about developing a deep, authentic understanding of who you are and what you bring to the table as a human being. Because when you have that kind of self-awareness and confidence in your own value, it becomes so much easier to communicate that value to others and position yourself for success in any field or industry.

Now, this process of self-discovery and skill identification isn't always easy - and it's definitely not a one-time thing. As you grow and evolve throughout your life and career, your unique strengths and edges are likely to shift and change as well. And that's okay - in fact, it's great! The key is to stay curious and open to learning new things about yourself, and to be proactive in seeking out opportunities to build and showcase your evolving skillset.

Because in this world that's changing so rapidly and profoundly, the most valuable skill of all might just be the ability to constantly learn, grow, and adapt. And by cultivating a deep understanding of your own unique strengths and abilities, and intentionally developing them over time, you'll be setting yourself up for a lifetime of personal and professional growth - no matter what the future holds.

So take the time to discover your human edge. Reflect on your experiences, pay attention to your natural strengths and interests, and seek out tools and frameworks that can help you gain a deeper understanding of what makes you uniquely valuable. Because in the age of AI and automation, it's the distinctly human skills and qualities that will set you apart and help you thrive - and the more you can understand and cultivate those, the better positioned you'll be for whatever comes next.

6.2. Developing A Growth Mindset

So, what exactly is a growth mindset? Simply put, it's the belief that your abilities and intelligence can be developed and improved over time through hard work, dedication, and a willingness to learn from your experiences - both the successes and the failures. It's about embracing challenges as opportunities for growth, rather than shying away from them out of fear or self-doubt.

That might sound easier said than done - after all, nobody likes to fail or feel like they're not good enough. But here's the thing: failure is an inevitable part of the learning process. It's how we learn what doesn't work, and how we figure out what to do differently next time. And if we can learn to see failure not as a reflection of our worth or ability, but as a valuable source of feedback and insight, we open ourselves up to a whole world of growth and possibility.

Take the famous inventor Thomas Edison, for example. Did you know that he allegedly failed over 1,000 times before successfully inventing the light bulb? When asked about his failures, he reportedly said, "I didn't fail 1,000 times. The light bulb was an invention with 1,000 steps." Talk about a growth mindset! Edison saw each failure as a necessary step on the path to success, and he never let setbacks or obstacles discourage him from pursuing his vision.

But developing a growth mindset isn't just about being resilient in the face of failure - it's also about actively seeking out new challenges and opportunities to learn and grow. It's about cultivating a sense of curiosity and wonder about the world around you, and always being open to new ideas and perspectives.

One way to do this is by setting achievable goals for yourself and tracking your progress over time. Whether it's learning a new skill, taking on a new project at work, or even just reading a book on a topic that interests you, having a clear sense of what you want to achieve and how you're going to get there can be a powerful motivator. And by celebrating your successes along the way, no matter how small, you reinforce the idea that growth and progress are possible with effort and dedication.

But of course, growth and learning don't happen in a vacuum - they require feedback, support, and guidance from others. That's why seeking out constructive criticism and advice from mentors, peers, and experts is so important. It can be scary to put yourself out there and ask for feedback, especially if you're not used to it. But by learning to see feedback as a gift, rather than a threat, you open yourself up to a whole world of insights and opportunities for improvement.

Now, developing a growth mindset is one thing - but what about actually putting it into practice in your life and work? That's where the idea of interdisciplinary skill integration comes in. In a world that's changing as rapidly as ours is, it's not enough to just be good at one thing - you need to be able to combine and apply your skills in new and creative ways.

Take the field of data science, for example. On its own, technical expertise in programming, statistics, and machine learning is certainly valuable. But when you combine that expertise with skills like creativity, communication, and ethical reasoning, you open up a whole new world of possibilities. You're not just crunching numbers anymore - you're telling stories, solving complex problems, and making decisions that have real-world impacts on people's lives.

Or think about the world of business and entrepreneurship. Sure, having a great product or service is important - but it's not enough on its own. To really succeed, you need to be able to combine your technical skills with things like emotional intelligence, leadership, and complex communication. You need to be able to build and motivate teams, negotiate with stakeholders, and navigate the complex social and political dynamics of any organization.

The point is, in the age of AI and beyond, the most successful and impactful individuals will be those who can integrate and apply their skills across multiple domains and contexts. They'll be the ones who can think creatively, reason ethically, communicate effectively, and adapt to new challenges and opportunities as they arise.

And that's where having a growth mindset comes in. Because if you believe that your abilities and intelligence can be developed over time, and you're willing to put in the hard work and effort required to keep learning and growing, there's no limit to what you can achieve. You're not just acquiring a fixed set of skills - you're building a foundation for lifelong learning and adaptation.

So whether you're a student just starting out on your educational journey, or a seasoned professional looking to stay relevant and impactful in a rapidly changing world, remember this: your mindset matters. Embrace challenges, seek out feedback, and never stop learning and growing. Because in the end, it's not just about what you know - it's about what you're capable of becoming.

6.3. Crafting Your AI-Age Personal Brand

Alright, so you've done the hard work of figuring out your unique human strengths and superpowers and potentially acquired a growth mindset. Congrats! That's a huge step towards thriving in the AI age. But now comes the next challenge - how do you take those awesome qualities and package them up in a way that really resonates with the world around you?

Enter the art of personal branding. Now, before you start having flashbacks to cheesy self-help books or cringe-worthy LinkedIn profiles, hear me out. Crafting a compelling personal brand is not about being fake or inauthentic. It's about taking the time to thoughtfully position and communicate your unique value in a way that cuts through the noise and makes people sit up and take notice.

And in the age of AI and automation, this is going to be more important than ever. With machines taking over more and more tasks and roles, the skills and qualities that will really shine are the ones that are distinctly human - things like creativity, empathy, strategic thinking, and the ability to connect with others on a deep level. By building a personal brand that showcases and celebrates those AI-complementary abilities, you'll be setting yourself up to stand out and thrive no matter what the future holds.

So where do you start? Well, one key element of a strong personal brand is having a clear and compelling mission statement. This is basically a short, punchy summary of who you are, what you stand for, and how you're looking to make an impact in the world. It's like a north star that guides all of your branding and communication efforts and helps you stay focused and authentic in how you show up.

For example, let's say you're really passionate about using your creativity and storytelling skills to inspire social change. Your personal mission statement might be something like: "I use the power of words and imagination to spark empathy, action, and transformation around the issues that matter most." Boom - that's a powerful and memorable way to sum up what makes you tick, and how you're looking to use your unique strengths to make a difference.

Of course, a mission statement is just one piece of the personal branding puzzle. Another key aspect is creating an online presence that really showcases your AI-complementary skills and experiences. This could be anything from a sleek personal website or portfolio to a killer LinkedIn profile or social media presence. The key is to be intentional and strategic about how you're presenting yourself to the world and to make sure that everything you put out there is aligned with your unique human edge.

For example, let's say you're a marketing whiz with a knack for understanding consumer psychology and crafting compelling brand stories. You might create a personal website that highlights some of your best work in those areas, along with thought leadership pieces or case studies that demonstrate your expertise. You might also make sure your LinkedIn profile is optimized with keywords and descriptions that emphasize your skills in things like consumer insights, brand strategy, and storytelling. By creating a cohesive and compelling online brand, you'll be making it easy for people to understand and connect with the unique value you bring to the table.

Another important aspect of personal branding in the AI age is being intentional about the overall narrative and story you're telling about yourself. This is not about crafting some fake persona or trying to be something you're not. It's about taking the time to reflect on your journey, your experiences, and your unique perspective, and finding ways to weave those elements together into a cohesive and compelling story.

For example, let's say you're someone who started out in a totally different field, but through a series of unexpected twists and turns, discovered a passion for using your emotional intelligence and people skills to build strong teams and organizations. That's a powerful story of growth, adaptability, and discovering your true calling - and it's a story that can really resonate with others who might be going through similar journeys.

By crafting an authentic and compelling personal brand narrative, you'll be giving people a reason to root for you and feel invested in your success. You'll also be setting yourself apart from all the noise and competition out there - because let's face it, in a world where everyone's got a carefully curated Instagram feed and a shiny LinkedIn profile, it's the real, human stories that truly stand out.

So how do you actually go about building a strong personal brand in practice? Well, one great way to start is by looking at some successful examples of people who have really nailed the whole "human-first" branding thing. Take someone like Brené Brown, for example. She's a researcher and storyteller who has built an incredible brand around the power of vulnerability, courage, and authentic human connection. Through her books, her TED talks, and her overall presence, she's crafted a powerful and cohesive narrative about the importance of embracing our imperfections and showing up fully in life and work.

Or look at someone like Gary Vaynerchuk, the entrepreneur and internet personality who's built a massive following by consistently showcasing his hustle, his creativity, and his ability to connect with people on a real, human level. Sure, he's got a slick website and a killer social media game - but at the end of the day, it's his authenticity and his willingness to share his own journey that really sets him apart.

The point is, there are tons of great examples out there of people who have successfully crafted compelling personal brands that showcase their unique human strengths. And by studying those examples and reverse-engineering what makes them work so well, you can start to build your own roadmap for personal branding success.

But here's the key - at the end of the day, your personal brand is not about some surface-level, manufactured image or persona. It's about taking the time to really understand and articulate what makes you uniquely valuable as a human being, and then finding authentic and compelling ways to share that with the world. It's about being true to yourself, while also being strategic and intentional about how you're showing up and making an impact.

And in a world that's changing as rapidly and profoundly as ours is, that kind of authentic, human-first branding is going to be more important than ever. So don't be afraid to put yourself out there, to share your story and your unique value with the world. Because the AI age isn't just about the rise of machines - it's also about the rise of a new kind of human, one who's confident, creative, and ready to thrive no matter what the future holds.

6.4. Networking And Relationship Building

So you've done the hard work of discovering your unique human edge and crafting a killer personal brand around it. Congrats - you're already ahead of the game! But here's the thing - in today's hyper-connected, always-on world, having a strong personal brand is really just the first step. To truly thrive in the age of AI and automation, you've got to know how to build and leverage relationships like a pro.

I mean, think about it - as machines get smarter and more capable, they're taking over more and more of the routine, transactional tasks that used to make up so much of our day-to-day work. But you know what machines are still pretty terrible at? Building genuine, trust-based relationships with other humans. They might be able to crunch data and optimize processes like nobody's business, but when it comes to things like empathy, emotional intelligence, and the ability to connect with people on a real, personal level? That's all us, baby.

And make no mistake - those high-touch, deeply human skills are only going to become more valuable and in demand as the AI revolution accelerates. In a world where so much can be automated or optimized, the ability to build strong relationships, to communicate and collaborate effectively, and to inspire and persuade others is going to be the ultimate competitive advantage.

So how do you go about building those crucial relationships and growing your network in a way that feels authentic and meaningful? Well, there are a few key strategies that I think are especially important in the digital age.

First and foremost, focus on quality over quantity. In the age of social media and digital networking, it's easy to get caught up in the idea that more connections equals more success. But the truth is, having a huge network of shallow, transactional relationships is way less valuable than having a smaller group of deep, genuine connections with people who know, like, and trust you.

Instead of trying to rack up as many LinkedIn connections or Instagram followers as possible, focus on building real, meaningful relationships with the people who matter most to you and your goals. Take the time to really get to know them, to understand their needs and challenges, and to find ways to support and add value to their lives and careers.

One great way to do this is through good old-fashioned face-to-face networking. I know, I know - in the age of COVID and remote work, in-person networking might feel like a bit of a lost art. But there's still something so powerful about being in the same room as someone, looking them in the eye, and having a real, human conversation.

So even if you're working remotely or in a hybrid setup, make an effort to get out there and connect with people in person when you can. Attend industry events and conferences, join local meetup groups or professional associations, or just grab coffee or lunch with colleagues and contacts whenever possible. The more you can build those in-person relationships and showcase your unique human skills in a face-to-face setting, the more you'll stand out and thrive in an increasingly digital world.

Of course, virtual networking is also a hugely important piece of the puzzle these days. And there are tons of great ways to build meaning-ful relationships and showcase your skills and expertise online. One key strategy is to be strategic and intentional about the platforms and channels you use to connect with others.

For example, if you're looking to build relationships with potential employers or clients, LinkedIn is obviously going to be a crucial plat-form. But don't just set up a basic profile and call it a day - take the time to optimize your presence and showcase your unique value proposition.

Share thought leadership content that demonstrates your expertise and insights, engage with others' posts and articles in a thoughtful and authentic way, and proactively reach out to people in your network who you admire or want to learn from.

If you're in a more creative or visually-driven field, platforms like Instagram, YouTube, or even TikTok can also be great ways to build your brand and connect with others. The key is to be authentic and consistent in how you show up and to focus on creating content that really resonates with your target audience and showcases your unique human strengths.

Another important aspect of virtual networking is being proactive and generous in how you engage with others. Don't just sit back and wait for opportunities or connections to come to you - go out there and make them happen! Reach out to people you admire and ask for informational interviews or virtual coffee chats. Offer to help or support others in your network whenever you can, whether that's making an introduction, sharing a resource, or just being a friendly ear to bounce ideas off of.

The more you can show up as a valuable, generous, and authentic human in your virtual interactions, the more you'll build a reputation as someone worth connecting with and investing in. And in a world where so much of our professional and personal lives are happening online, that kind of digital reputation and social capital is only going to become more important.

Of course, building strong relationships and growing your network is not always easy - and it's definitely not a one-time thing. Just like any other important skill or asset, your social capital requires ongoing investment and cultivation. But the payoffs - both personally and professionally - can be truly incredible.

I mean, think about some of the most successful and influential people throughout history. Whether it's leaders like Nelson Mandela or entrepreneurs like Oprah Winfrey, so much of their impact and success has come down to their ability to build and leverage strong relationships with others. They understood that in a world full of complex challenges and opportunities, the power of human connection and collaboration is the ultimate force multiplier.

And that's a lesson that I think is only going to become more important in the age of AI and automation. As machines take over more and more of the routine and repetitive tasks in our lives and work, it's the deep human skills of empathy, communication, and relationship-building that will truly set us apart and help us thrive.

Don't underestimate the power of your own human edge when it comes to networking and building relationships. Focus on quality over quantity, be authentic and generous in how you show up, and always look for ways to add value and support to those around you. Because in a world that's changing faster than ever, it's the strength of our human connections that will ultimately determine our success and impact.

6.5. Marketing Your Human Skills

Alright, you've identified your unique human strengths, you've crafted a killer personal brand around them, and you've been building relationships and growing your network like a pro. You're basically a superhero of the AI age at this point. But there's one more crucial step in positioning yourself for success - and that's learning how to effectively market and showcase your human skills to the world.

I mean, let's face it - in today's hypercompetitive job market, it's not enough to just have a strong set of skills and experiences. You've got to know how to package and present them in a way that really grabs people's attention and makes them sit up and take notice. And when it comes to highlighting your distinctly human abilities in a world of automation and AI, that's more important than ever.

Let's start with the basics - your resume, portfolio, and online profiles. These are often the first things that potential employers, clients, or collaborators will see when they're considering you for an opportunity, so it's crucial that they're optimized to showcase your unique human strengths.

One key strategy here is to focus on highlighting your most AI-resistant skills and experiences. So instead of just listing off a bunch of technical competencies or software proficiencies that could easily be automated, focus on emphasizing things like your creativity, your problem-solving abilities, your emotional intelligence, or your knack for communication and collaboration.

For example, let's say you're a graphic designer looking for your next gig. Instead of just listing off all the design software you know how to use (which, let's be real, is pretty much expected at this point), focus on highlighting the unique creative vision and strategic thinking you bring to your work. Talk about how you've used your design skills to solve complex business problems, or how you've collaborated with cross-functional teams to create truly innovative and impactful campaigns.

Or let's say you're a recent grad looking to break into the world of marketing. Instead of just listing off your degree and coursework (which, again, is kind of a given), focus on showcasing real-world projects and experiences that demonstrate your unique human skills. Maybe you led a student organization and had to rally a team around a shared vision, or maybe you volunteered for a local nonprofit and used your communication skills to create compelling content and campaigns. Whatever it is, make sure you're highlighting the distinctly human abilities that set you apart.

But of course, your resume and online presence are just the first step. Once you've landed that interview or that networking coffee chat, it's time to really put your human skills to work and showcase the unique value you bring to the table.

One key strategy here is to focus on storytelling and specific examples. Instead of just claiming to be a "strong communicator" or a "creative problem-solver," share concrete stories and anecdotes that illustrate those abilities in action. Talk about a time when you had to navigate a complex social situation at work, or when you came up with an out-of-the-box solution to a seemingly impossible challenge.

The more specific and vivid you can be in your examples, the more memorable and impactful your pitch will be. And don't be afraid to show a little vulnerability or humility in the process - after all, part of what makes us human is our ability to learn, grow, and overcome challenges along the way.

Another important aspect of marketing your human skills is being proactive and persistent in your efforts. Don't just wait around for opportunities to fall into your lap - go out there and create them yourself! Reach out to people in your network whom you admire and ask for informational interviews or advice. Attend industry events and conferences and make a point of connecting with speakers and attendees who share your interests and values. Volunteer for projects or initiatives that allow you to stretch your skills and make a real impact.

The more you can take control of your own career narrative and proactively showcase your unique human strengths, the more successful and fulfilled you'll be in the long run. And don't be afraid to iterate and experiment along the way - part of the beauty of being human is our ability to adapt, learn, and grow over time.

Now, I know that all of this marketing and self-promotion stuff can sometimes feel a little icky or inauthentic. And believe me, I get it - nobody wants to feel like they're constantly having to sell themselves or prove their worth. But here's the thing - in a world that's changing as rapidly and profoundly as ours is, being able to effectively communicate your value and differentiate yourself from the crowd is not just a nice-to-have, it's a necessity.

And when you really think about it, marketing your human skills is not about being fake or salesy - it's about being authentic and purposeful in how you show up in the world. It's about taking the time to reflect on your unique strengths and experiences, and then finding ways to share those with others in a way that's genuine, relevant, and impactful.

Because at the end of the day, your human skills are your ultimate competitive advantage - they're what sets you apart from the machines and algorithms, and what enables you to create real value and meaning in your life and work. So don't be afraid to own them, to celebrate them, and to share them with the world. The age of AI may be here, but it's our distinctly human abilities that will ultimately determine our success and fulfillment in the years ahead.

CHAPTER 7

HUMAN-AI COLLABORATION MODELS

We've spent a lot of time talking about the incredible capabilities of AI, and how we can cultivate our own uniquely human skills to stay relevant and thrive in a world where machines are getting smarter every day. But what if I told you that the real magic happens when we bring those two things together? When we stop seeing AI as a rival or a replacement, and start thinking about how we can collaborate with these powerful tools to achieve things that neither of us could do alone?

Welcome to the exciting world of human-AI collaboration - the key to unlocking a whole new level of innovation, creativity, and problem-solving in the age of artificial intelligence.

In this chapter, we're going to explore some of the most cutting-edge frameworks and models for making this collaboration happen. We'll look at how humans and AI can form "teams of rivals" - working together in a spirit of friendly competition to tackle complex challenges from multiple angles. We'll borrow principles from game theory and cognitive science to understand how we can create feedback loops and joint systems that amplify the strengths of both human and machine intelligence.

We'll also dive into some of the latest advances in human-computer interaction, like "centaur" models that create tight couplings between human operators and AI systems. Imagine being able to guide and train a machine learning algorithm in real time, using your own expertise and intuition to shape its outputs. Or picture an AI system that can explain its reasoning and decision-making process in a way that's transparent and understandable to its human collaborators.

From there, we'll explore how these collaboration models can be applied to real-world decision-making scenarios - from business strategy and policy-making to medical diagnosis and scientific discovery. We'll look at how AI can provide powerful tools for simulation, prediction, and optimization, while humans bring their own judgment, context awareness, and ethical sensibilities to the table. The result? A kind of "augmented intelligence" that's more than the sum of its parts.

But of course, "with great power comes great responsibility". As AI systems become more advanced and autonomous, we must have robust governance frameworks in place to ensure that they're being developed and deployed in a way that aligns with human values and priorities. We'll explore some of the latest thinking on AI ethics and safety, and look at how we can create technical and regulatory safeguards to keep humans in the loop and in control.

Now, I know some of this might sound a bit daunting at first - like something out of a science fiction novel or a Silicon Valley tech demo. But the truth is, these collaboration models are already being used in all sorts of fields and industries, from finance and healthcare to transportation and entertainment. And as AI continues to advance at a rapid pace, the opportunities for human-machine teamwork are only going to grow.

So whether you're a budding data scientist looking to build the next generation of intelligent systems, or a business leader trying to stay ahead of the curve in a rapidly changing market, this chapter is for you. By understanding the principles and practices of effective human-AI collaboration, you'll be positioning yourself at the forefront of one of the most transformative technological shifts of our time.

But more than that, you'll be tapping into something fundamentally human - our ability to learn, to adapt, to create, and to work together towards a common goal. Because in the end, the true potential of AI isn't about replacing humans, but about amplifying our own unique capabilities and helping us to achieve things that we never thought possible.

7.1. The Human-AI Team Of Rivals

Let's begin with diving into one of the most exciting and powerful models for human-AI collaboration - the idea of forming "teams of rivals" to tackle complex problems. Although you might wonder - rivals? Isn't the whole point of collaboration to work together in harmony? But hear me out, because this is where things start to get really interesting.

You see, when we talk about human-AI teams of rivals, we're not talking about a bitter competition or a zero-sum game. Instead, we're talking about a kind of friendly rivalry - a push and pull between two different forms of intelligence that ultimately makes both of them stronger.

To understand how this works, let's borrow a few concepts from the world of game theory. One of the key ideas here is the notion of iterative plays - the idea that complex problems are often solved not in one fell swoop, but through a series of back-and-forth exchanges between different players. Each player brings their own unique strengths and perspectives to the table, and through a process of trial and error, they gradually converge on a solution that's better than what either of them could have come up with on their own.

Now, in the case of human-AI teams, we can think of the human as one player and the AI as another. The human brings their own deep knowledge of the problem space, their ability to think creatively and laterally, and their understanding of the broader context and implications of the problem at hand. The AI, on the other hand, brings its raw computational power, its ability to crunch through vast amounts of data and identify patterns and insights that might be invisible to the human eye.

By working together in a kind of iterative feedback loop, the human and the AI can play to each other's strengths and compensate for each other's weaknesses. The human might start by defining the problem space and setting the initial parameters for the AI to work within. The AI then goes off and generates a whole bunch of potential solutions, using its algorithms and machine learning models to explore the space of possibilities.

But the process doesn't stop there. The human then comes back in and evaluates the AI's outputs, using their own judgment and domain expertise to identify the most promising ideas and filter out the ones that don't make sense. They might tweak the parameters or reframe the problem based on what they've learned, and then send the AI back to work on the next iteration.

Through this kind of back-and-forth loop, the human and the AI are able to converge on solutions that are both technically feasible and contextually appropriate. It's a kind of joint cognitive system, where the whole is greater than the sum of its parts.

Now, if this all sounds a bit abstract, let's look at some concrete examples of how this model is already being used in the real world. One of my favorite case studies comes from the world of drug discovery - a field where the sheer complexity of the problem space makes it incredibly difficult for either humans or AI to make progress on their own.

Back in 2020, a team of researchers from MIT and the pharmaceutical company Novo Nordisk decided to try a new approach. They created a human-AI team where the humans were responsible for defining the specific molecular properties they were looking for in a potential drug candidate, and the AI was responsible for generating novel chemical compounds that fit those criteria.

Through a series of iterative cycles, the team was able to identify several promising drug candidates that had never been seen before - compounds that the humans wouldn't have been able to dream up on their own, and that the AI wouldn't have been able to identify without the humans' guidance and feedback.

But the power of human-AI teams of rivals isn't limited to the realm of science and technology. We're also starting to see this model being used in fields like business strategy, policy-making, and even creative industries like music and art.

In the world of finance, for example, there are now AI systems that can analyze vast amounts of market data and identify patterns and trends that human analysts might miss. But rather than just blindly following the AI's recommendations, savvy investors are using these insights as a starting point for their own analysis and decision-making - a way to augment and enhance their own expertise rather than replace it.

And in the realm of music, there are now AI algorithms that can generate entire compositions in the style of famous musicians or genres. But rather than seeing this as a threat to human creativity, some artists are using these tools as a way to spark new ideas and push their own boundaries. They might take an AI-generated melody and use it as the basis for a new song, or collaborate with an AI system to create a piece that blends both of their unique styles and sensibilities.

The point is, by forming teams of rivals with AI, we're not giving up our own agency or creativity as humans. We're simply recognizing that there are some things that machines are really good at, and some things that humans are really good at - and by bringing those two forms of intelligence together, we can achieve things that neither of us could do alone.

Of course, this isn't to say that human-AI collaboration is always easy or straightforward. There are certainly challenges and pitfalls to be aware of, from the risk of bias and errors in the AI's outputs to the need for clear communication and trust between human and machine partners.

But overall, the potential benefits of this model are simply too great to ignore. By learning to work alongside AI as teammates and collaborators, we're not just staying relevant in a changing world - we're actively shaping the future and pushing the boundaries of what's possible.

So the next time you find yourself facing a complex problem or challenge, don't just rely on your own human brainpower to solve it. Consider bringing in an AI rival to help you out - and get ready to be amazed at what you can achieve together.

7.2. Human-AI Centaur Models

Now let's talk about one of the coolest and most cutting-edge models for human-AI collaboration out there - the "centaur" model. And here is where you might think - centaurs? Like those half-human, half-horse creatures from Greek mythology? Well, not exactly. But the idea behind the name is actually pretty similar.

You see, in the world of human-computer interaction and cybernetics (which is just a fancy way of saying the study of how humans and machines interact), the centaur model is all about creating super tight, symbiotic relationships between humans and AI systems. It's about blurring the lines between where human intelligence ends and machine intelligence begins and creating a kind of hybrid intelligence that's more than the sum of its parts.

To understand how this works, let's break it down into a couple of key concepts. First up, we have the idea of human-guided machine learning. This is where humans play a really active role in training and shaping AI algorithms, rather than just letting them loose on a bunch of data and hoping for the best.

For example, let's say you're trying to build an AI system that can identify different species of birds in photographs. You could just feed it a massive dataset of bird images and let it figure out the patterns and features on its own. But with human-guided machine learning, you might start by having a human expert label a smaller subset of those images, identifying key features like beak shape, feather color, and wing patterns that distinguish different species from each other.

Then, as the AI starts to learn and make its own predictions, the human expert can step in and provide feedback - highlighting where the AI got things right, and pointing out where it made mistakes or missed important details. Over time, through this kind of iterative feedback loop, the AI can start to internalize the human's expertise and intuition and become much more accurate and reliable in its predictions.

But the centaur model isn't just about humans teaching AI - it's also about AI teaching humans. This is where we get into the realm of human-comprehensible AI - the idea that AI systems should be designed to expose their reasoning and decision-making processes in a way that's understandable and interpretable to their human partners.

For example, let's say you're working with an AI system that's helping to diagnose medical conditions based on patient data. Instead of just spitting out a diagnosis and leaving it at that, a human-comprehensible AI might also provide a clear explanation of how it arrived at that conclusion - highlighting the specific symptoms, test results, and risk factors that it weighed in its analysis.

This kind of transparency and interpretability is really important for building trust and accountability in human-AI partnerships. It allows humans to double-check the AI's reasoning, catch potential errors or biases, and make more informed decisions based on the AI's outputs.

Another key aspect of the centaur model is the idea of interactive retrofits - basically, finding ways to integrate AI capabilities into existing human workflows and interfaces, rather than starting from scratch with totally new systems.

For example, let's say you're a financial analyst who's used to working with a particular spreadsheet tool to analyze market data and make investment recommendations. Instead of trying to replace that tool entirely with some kind of black-box AI system, an interactive retrofit might involve building AI-powered features and insights directly into the spreadsheet interface itself.

So as you're working through your usual analysis, the AI might be running in the background, crunching through massive datasets and surfacing key patterns and anomalies that you might have missed on your own. It might even suggest alternative scenarios or risk factors to consider, based on its own predictive models and simulations.

The point is, by integrating AI capabilities into the tools and work-flows that humans are already comfortable with, we can create a much more seamless and intuitive collaboration experience - one where the human and the machine are truly working together as a single, cohesive unit.

Now, if you're thinking that all of this sounds a bit like science fiction - well, you're not entirely wrong. The idea of human-AI centaurs has been around for decades in the realm of speculative fiction and futurism. But the truth is, we're already starting to see real-world examples of this kind of collaboration in action.

One of my favorite examples comes from the world of chess - a domain where humans and machines have been competing and collaborating for decades. Back in the late 90s, when IBM's Deep Blue computer first beat world champion Garry Kasparov in a highly publicized match, many people saw it as a sign that machines were finally surpassing human intelligence in this domain.

But what happened next was even more interesting. In the years that followed, a new kind of chess player emerged - the so-called "centaur" player, who used a combination of human intuition and machine calculation to achieve superhuman levels of play. These players would use AI chess engines to analyze positions and generate potential moves, but they would also rely on their own strategic judgment and creative thinking to make the final decisions.

And the results were stunning. In 2005, a team of centaur players actually beat a team of grandmasters and a team of top-ranked AI chess engines in a tournament - a feat that would have been unthinkable just a few years earlier.

Of course, chess is just one domain where human-AI centaurs are starting to make their mark. We're also seeing this kind of collaboration emerge in fields like medical diagnosis, scientific research, and even creative industries like music and art.

The point is that the centaur model represents a powerful new paradigm for human-machine collaboration - one that doesn't just treat AI as a tool or a servant, but as a true partner and collaborator in the pursuit of knowledge and innovation. And while there are certainly challenges and risks to be navigated along the way, the potential benefits are simply too great to ignore.

So as you navigate your own path in this brave new world of AI, keep the centaur model in mind. Embrace the opportunity to work alongside these incredible machines, and to create something truly extraordinary together. Because in the end, that's what the future of intelligence is all about - not human versus machine, but human and machine, working together in pursuit of a common goal.

7.3. Human-AI Decision Augmentation

Okay, so we've talked about the power of human-AI collaboration and how the centaur model can help us achieve incredible things by combining the strengths of both human and machine intelligence. But now I want to dive into a specific area where this kind of collaboration is already having a huge impact, and that's in the realm of decision-making.

You see, whether you're a business leader trying to navigate a complex market, a policymaker grappling with a thorny social issue, or a doctor trying to diagnose a tricky medical case, making good decisions is hard. There are often a lot of variables to consider, a lot of uncertainty and ambiguity to navigate, and a lot of competing priorities and stakeholder interests to balance.

But what if you had an AI partner that could help you cut through that complexity and make better decisions faster? That's the idea behind human-AI decision augmentation - using AI systems to enhance and elevate human decision-making processes, while still keeping humans firmly in the driver's seat.

So how does this actually work in practice? Well, one key concept is the idea of probabilistic scenario simulations. Basically, this is where an AI system can take a complex problem or decision space and generate a whole bunch of potential scenarios or outcomes, each with its own probability and risk profile.

For example, let's say you're a business leader trying to decide whether to launch a new product line. You could feed all of your market research, customer data, and financial projections into an AI system, and it could generate a range of potential scenarios - from the best-case scenario where the product is a huge hit and generates massive profits, to the worst-case scenario where it flops and costs the company millions.

But the AI doesn't just spit out these scenarios and call it a day - it also provides a clear picture of the likelihood of each scenario, based on the data and assumptions you've provided. So you might see that there's a 60% chance of the product being moderately successful, a 20% chance of it being a huge hit, and a 20% chance of it being a total bust.

Now, this kind of probabilistic modeling is incredibly powerful, because it allows you to make more informed and confident decisions based on data rather than just gut instinct or guesswork. But of course, it's not a silver bullet - there are always going to be factors and considerations that the AI might miss or get wrong.

And that's where the human side of the equation comes in. By injecting your own pragmatic oversight and judgment calls into the decision-making process, you can help to validate and refine the AI's outputs, and ensure that they align with your own strategic priorities and values.

For example, let's say the AI predicts that there's a 70% chance of the new product being successful if you price it at $100 per unit. But based on your own industry experience and customer knowledge, you might know that the price point is actually way too high for your target market. So you can adjust the assumptions and parameters accordingly, and see how that changes the AI's predictions and recommendations.

This kind of human-AI decision augmentation is already starting to transform a wide range of industries and domains. In healthcare, for example, there are now AI systems that can analyze patient data and generate personalized treatment recommendations for doctors to consider. But rather than just blindly following the AI's advice, doctors can use their own clinical judgment and patient knowledge to validate and refine those recommendations, ensuring that they're tailored to each individual patient's needs and preferences.

Similarly, in the world of finance, there are now AI-powered trading systems that can analyze vast amounts of market data and generate trading strategies in real time. But rather than just letting the machines run wild, human traders and portfolio managers can use their own market intuition and risk tolerance to guide and constrain the AI's actions, ensuring that they align with the firm's overall investment philosophy and goals.

And in the realm of policymaking, there are now AI systems that can help governments and organizations simulate the potential impacts of different policy options and interventions, from tax changes to public health campaigns. But rather than just relying on the AI's outputs blindly, policymakers can use their own domain expertise and stakeholder input to refine and adjust those simulations, ensuring that they're grounded in real-world practicalities and priorities.

Now, of course, this kind of human-AI decision augmentation doesn't come without its challenges and risks. There are always going to be questions around bias, transparency, and accountability when it comes to using AI in high-stakes decision-making contexts. And there's always the risk of over-reliance on the AI's outputs, or of humans getting complacent and letting the machines do too much of the heavy lifting.

But overall, the potential benefits of this approach are simply too great to ignore. By leveraging the power of AI to help us navigate complexity and uncertainty, while still retaining our own human judgment and values, we can make better decisions faster - decisions that are more informed, more confident, and more aligned with our strategic goals and priorities.

And the great thing is, this isn't just some far-off future vision - it's already happening all around us, in a wide range of industries and domains. From healthcare and finance to marketing and supply chain management, organizations are already starting to harness the power of human-AI decision augmentation to drive better outcomes and stay ahead of the curve.

So if you're a student or young professional just starting out in your career, keep this in mind as you navigate your own path forward. The ability to work effectively with AI in a decision-making context is rapidly becoming a key skill and differentiator in today's job market - one that can open up a whole new world of opportunities and possibilities.

But more than that, it's a chance to be part of something truly transformative - to help shape the future of decision-making itself, and to create a world where human and machine intelligence work together seamlessly to solve our biggest challenges and achieve our boldest ambitions. And that, my friends, is a future worth getting excited about.

7.4. Governance And Human Control Of AI Systems

As we continue to push the boundaries of what's possible with artificial intelligence, it's also important that we take a step back and think carefully about how we're going to govern and control these incredibly powerful systems.

You see, as AI gets smarter and more capable, there's a very real risk that it could start to operate in ways that are harmful or misaligned with human values and priorities. We've all seen sci-fi movies where the machines take over and start wreaking havoc, and while that might still be a bit far-fetched, there are certainly more immediate and realistic concerns around things like bias, transparency, and accountability in AI systems.

So how do we make sure that we're developing and deploying AI in a way that's safe, responsible, and aligned with human interests? Well, that's where the idea of AI governance and human control comes in.

At a high level, AI governance is all about putting in place frameworks, policies, and processes to ensure that AI is being developed and used in an ethical and responsible way. This can include things like establishing clear guidelines and best practices for AI development, creating oversight and accountability mechanisms to monitor AI systems in action, and fostering ongoing dialogue and collaboration between AI developers, domain experts, policymakers, and the broader public.

One key aspect of AI governance is the idea of "human-in-the-loop" control - basically, ensuring that humans always have the ability to intervene and adjust AI systems as needed, rather than just letting them run on autopilot. This can take a lot of different forms, depending on the specific context and application.

For example, in the realm of autonomous vehicles, there's a concept called "enclaving" - essentially, creating designated zones or scenarios where the AI system is allowed to operate autonomously, but with clear boundaries and failsafes in place to ensure that it stays within safe and acceptable parameters. So you might have an autonomous car that can navigate city streets on its own but with hard-coded rules around things like speed limits, pedestrian safety, and emergency stopping.

Similarly, in the world of healthcare, there's a lot of interest in using AI to help with things like diagnosis, treatment planning, and drug discovery. But rather than just letting the AI make all the decisions on its own, many experts advocate for a "human-in-the-loop" approach where doctors and other medical professionals are always involved in reviewing and validating the AI's outputs before they're put into practice.

Another important concept in AI governance is the idea of "corrigibility" - basically, the ability to correct or adjust an AI system if it starts to behave in unintended or harmful ways. This is especially important as AI systems become more complex and opaque, and it becomes harder for humans to understand exactly how they're making decisions or arriving at certain outputs.

One way to build corrigibility into AI systems is through the use of "tripwires" - essentially, hard-coded triggers or thresholds that will automatically shut down or adjust the AI if certain conditions are met. For example, you might have an AI trading system that's designed to optimize financial returns, but with tripwires in place to prevent it from engaging in illegal or unethical practices like insider trading or market manipulation.

But of course, technical solutions like enclaving and tripwires are only part of the equation when it comes to AI governance. There's also a huge role for policy, regulation, and institutional oversight in ensuring that AI is being developed and deployed responsibly.

One emerging best practice in this regard is the creation of AI ethics boards or review committees - basically, groups of experts and stakeholders from different domains who are responsible for overseeing the development and deployment of AI systems within a particular organization or industry. These boards can help to ensure that AI is being developed in line with established ethical principles and guidelines, and can provide a forum for ongoing dialogue and debate around the social and political implications of AI.

Another important policy lever is the idea of human approval pipelines - essentially, requiring that certain high-stakes AI decisions or outputs go through a human review and approval process before they're put into action. For example, you might require that any AI-generated medical diagnoses or treatment plans be reviewed and signed off by a licensed physician before they're given to patients.

Now, I know that all of this talk of governance and control might sound a bit boring and bureaucratic compared to the exciting possibilities of human-AI collaboration. But the truth is, getting this stuff right is absolutely essential if we want to realize the full potential of AI while also mitigating its risks and downsides.

And the stakes are only going to get higher as AI continues to advance and become more integrated into every aspect of our lives. I mean, just think about some of the incredible AI breakthroughs that have happened in recent years - things like AlphaFold cracking the protein folding problem, or GPT-3 generating shockingly coherent and creative language outputs.

As these systems become more powerful and capable, the need for robust governance and human control will only become more urgent. Because at the end of the day, we want to make sure that we're the ones steering the ship, not the machines.

And that's where you come in, as the next generation of leaders, innovators, and changemakers. By learning about these issues now and thinking critically about the ethical and social implications of AI, you can help shape a future where human and machine intelligence work together in a way that's safe, responsible, and aligned with our deepest values and aspirations.

So don't be afraid to ask the tough questions, to challenge assumptions, and to advocate for the kind of AI governance and control that you believe in. Because the choices we make now will have profound consequences for the future of humanity - and it's up to all of us to make sure that we get it right.

7.5. Data Labelling

Let's talk about one of the unsung heroes of the AI world - the crucial process of data labelling. Even though "Data labelling" sounds about as exciting as watching paint dry, trust me, this is where the magic happens. This is where humans and machines come together to create something greater than the sum of their parts.

So, what exactly is data labelling, and why is it so important? Well, imagine you're trying to teach a machine to recognize different types of animals. You can't just show it a bunch of pictures and expect it to figure it out on its own. You need to give it some guidance, some context, some way of understanding what it's looking at. That's where data labelling comes in.

Essentially, data labelling is the process of attaching meaningful tags or categories to raw data, like images, text, or audio files. It's about taking all that unstructured information and giving it some structure, some meaning, that a machine can understand and learn from. And here's the kicker - it's not something that machines can do on their own. It requires human intelligence, human judgment, and human expertise.

Think about it - when you look at a picture of a cat, you don't just see a bunch of pixels. You see a furry, four-legged creature with whiskers and a tail. You understand the concept of "cat" based on your own experiences and knowledge of the world. And that's the kind of contextual understanding that machines need to learn from.

So when a human labels an image of a cat as "cat," they're not just stating the obvious. They're providing a machine with a valuable piece of information that it can use to recognize other cats in the future. And the more high-quality, diverse, and representative data labels a machine has to learn from, the better it gets at understanding and interpreting the world around it.

But here's the thing - data labelling is not a one-and-done kind of job. It's an ongoing process that requires constant human input and oversight. As machines learn and improve, they need to be fed new and more complex data to keep expanding their knowledge and capabilities. And that data needs to be carefully curated and labelled by humans to ensure its accuracy and relevance.

In fact, data labelling has become a crucial part of the AI development process, with entire teams and even companies dedicated to providing high-quality data labelling services. It's a bit like the quality control department of the AI world - without accurate and reliable data labels, even the most advanced algorithms and models will fall short.

And the importance of data labelling goes beyond just teaching machines to recognize cats and dogs. It's being used in everything from self-driving cars and medical diagnosis to fraud detection and personalized recommendations. The applications are endless, but they all rely on the foundation of human-labelled data to function effectively.

For example, let's say you're developing an AI system to help doctors identify cancerous tumors in medical images. You can't just feed the machine a bunch of unlabeled scans and hope for the best. You need expert radiologists to carefully label each image, identifying the precise location and characteristics of any tumors or abnormalities. Only then can the machine learn to recognize those patterns on its own and assist doctors in making accurate diagnoses.

Or take the example of natural language processing, which we talked about in the previous section. In order for machines to understand and interpret human language effectively, they need to be trained on vast amounts of labelled text data. That means humans have to go through and tag things like parts of speech, named entities, sentiment, and intent - all the nuances and complexities that make human language so rich and expressive.

It's a bit like teaching a child to read - you start with the basics, like pointing out letters and sounding out simple words. But as the child grows and learns, you introduce more advanced concepts like grammar, context, and figurative language. And all along the way, you're providing guidance and feedback to help them understand and apply what they've learned.

The same is true for machines - they need that human guidance and feedback to learn and grow. And that's why data labelling is such an essential part of human-AI collaboration. It's not just about providing raw fuel for algorithms - it's about imbuing machines with human knowledge, human judgment, and human context.

In a way, data labelling is like the secret handshake between humans and machines - a way of communicating and collaborating that allows us to build something greater than either of us could achieve alone. And as the field of AI continues to evolve and mature, the role of human data labellers will only become more important and more valuable.

So if you're looking for a way to make your mark in the world of AI, don't overlook the power of data labelling. Whether you're an aspiring developer, a domain expert, or just someone with a keen eye for detail, there's a place for you in this crucial process. Because at the end of the day, it's not just about the algorithms or the models - it's about the human intelligence and expertise that brings them to life.

So let's raise a glass (or a well-labelled dataset) to the unsung heroes of the AI world - the human data labellers who make it all possible. Without you, the machines would be lost - and the future of AI would be a lot less bright.

CHAPTER 8

ETHICAL AI DEVELOPMENT

Wow, we've covered a lot of ground in this book so far, exploring the incredible possibilities of human-AI collaboration and how we can cultivate our own unique skills and strengths to thrive in an age of artificial intelligence.

But as we've seen, the greater the influence, the higher the accountability. And when it comes to developing and deploying AI systems that could potentially reshape every aspect of our lives, the stakes couldn't be higher.

That's why in this last chapter, we're going to be tackling one of the most important and challenging topics in the field of AI today: how to ensure that these incredibly powerful technologies are being developed and used in an ethical, responsible, and trustworthy way.

Now, I know that the word "ethics" might conjure up images of dry philosophy lectures or abstract moral dilemmas. But the truth is, the ethical considerations around AI are incredibly practical and relevant to all of our lives. We're talking about things like:

- How do we make sure that AI systems are fair and unbiased, and don't discriminate against certain groups of people?

- How do we protect people's privacy and personal data in a world where AI is constantly collecting and analyzing information about us?

- How do we ensure that AI is being used in ways that are transparent and accountable, and that humans can always understand and control what the machines are doing?

- And perhaps most importantly, how do we ensure that the development and deployment of AI are always aligned with human values and interests, and not just the pursuit of raw intelligence or efficiency?

These are the kinds of questions that we'll be grappling with in this chapter, as we explore the emerging field of ethical AI development. And make no mistake - these are not just abstract or theoretical concerns. The choices we make around these issues will have profound consequences for the future of humanity, and it's up to all of us to get it right.

So what does ethical AI development actually look like in practice? Well, as we'll see, it's about a lot more than just slapping a few rules or guidelines onto an AI system and calling it a day. It's about fundamentally rethinking the way we approach the development and deployment of these technologies, from the ground up.

For starters, it means aligning AI systems with human values and ethics from the very beginning, using techniques like value learning and inverse reinforcement learning to ensure that the machines are pursuing goals and objectives that are consistent with our own moral principles. It means hardwiring things like human rights, fairness, and accountability into the very architecture of these systems so that they're not just an afterthought or a bolt-on feature.

It also means adopting a more holistic and proactive approach to responsible AI development, one that considers the full lifecycle of these technologies from initial problem formulation all the way through to deployment and monitoring. This includes things like rigorously testing AI systems for potential biases or negative impacts, conducting regular audits and assessments to ensure that they're functioning as intended, and having clear processes in place for human oversight and control.

But perhaps most importantly, ethical AI development is about creating a culture and a community around these issues - one that brings together experts and stakeholders from all different domains to collaborate, share best practices, and hold each other accountable. It's about recognizing that the development of AI is not just a technical challenge, but a deeply social and political one as well, and that we all have a role to play in shaping its future.

So whether you're a computer science major looking to build the next generation of AI systems, a social science student exploring the societal implications of these technologies, or just someone who cares deeply about the future of humanity - this chapter is for you. By understanding the key principles and practices of ethical AI development, and by starting to grapple with these issues now, you can help to create a future where artificial intelligence truly benefits everyone, and not just a select few.

But we can't stop at just avoiding harm or mitigating risks - we also need to be proactive in shaping AI systems that actively promote human flourishing and well-being. Imagine AI that doesn't just play by the rules, but actually helps us to be more creative, more empathetic, and more fulfilled as individuals and as a society. Imagine AI that helps us to solve some of the greatest challenges facing humanity today, from disease and poverty to climate change and beyond.

That's the kind of AI that I believe is worth fighting for - and it all starts with approaching the development of these technologies with a deep sense of ethics, responsibility, and care. So let's roll up our sleeves and get to work - the future is waiting for us to build it, one line of code and one difficult conversation at a time.

8.1. Aligning AI With Human Values

First of all, let's discuss one of the most critical and challenging aspects of ethical AI development - the question of how we ensure that these incredibly powerful systems remain aligned with human values and ethics, even as they become more and more capable.

It's a tricky problem because when you think about it, values and ethics are inherently human concepts. They're shaped by our cultures, our histories, our individual experiences, and beliefs. And they're often messy, context-dependent, and open to interpretation.

So how do we take something as complex and nuanced as human morality and translate that into the rigid, binary language of computers and algorithms? How do we create AI systems that don't just blindly pursue some narrow objective, but actually understand and care about things like fairness, empathy, and the greater good?

Well, as it turns out, there are a few key concepts and approaches that researchers and developers are exploring to try to crack this problem. One of the big ones is the idea of value learning - essentially, creating AI systems that can observe human behavior and preferences and then infer the underlying values and principles that drive those actions.

For example, let's say you have an AI system that's designed to help manage a city's traffic flow. You could have it watch how human traffic controllers make decisions in different situations - when they choose to turn signals red or green, when they prioritize certain vehicles over others, and so on. And over time, the AI could start to pick up on the implicit values that guide those decisions - things like minimizing accidents, reducing congestion, and ensuring fair access for all drivers.

Another approach that's gaining a lot of attention is inverse reinforcement learning. The basic idea here is that instead of trying to explicitly program a set of values or ethical principles into an AI system, you let it figure out those values for itself by observing the behavior of experts in a particular domain.

So going back to the traffic example, you might have the AI watch how a panel of experienced traffic controllers handle different scenarios, and then use machine learning techniques to essentially reverse-engineer the reward function that those experts are optimizing for. And once the AI has internalized that reward function, it can start to make its own decisions in a way that aligns with the values and priorities of those human experts.

But of course, both of these approaches come with their own challenges and limitations. For one thing, there's the risk that the AI might pick up on biases or blind spots in the human behavior it's observing, and end up perpetuating those flaws in its own decision-making. There's also the question of how to handle situations where different human stakeholders might have conflicting values or priorities - like when a traffic controller has to choose between optimizing for speed or safety, for example.

That's where the idea of constitutional AI comes in - the notion of hardwiring certain fundamental values or principles into an AI system's core architecture, so that they serve as immutable constraints on its behavior. This could take the form of explicit rules or guidelines that the AI is required to follow, or it could be more like a set of weightings or priorities that shape how the AI makes trade-offs between different objectives.

One famous example of this idea comes from the science fiction writer Isaac Asimov, who proposed a set of "Three Laws of Robotics" that would serve as the ethical foundation for all artificial intelligence:

1. A robot may not injure a human being or, through inaction, allow a human being to come to harm.
2. A robot must obey the orders given it by human beings except where such orders would conflict with the First Law.
3. A robot must protect its own existence as long as such protection does not conflict with the First or Second Laws.

Now, these laws are obviously a bit simplistic and open to interpretation. And as Asimov himself explored in his stories, there are plenty of situations where they might break down or lead to unintended consequences. But the underlying idea - of baking certain non-negotiable human values into the core of an AI system - is a powerful one that's gaining a lot of traction in the field.

Some researchers are exploring ways to formalize this idea even further, by creating so-called "constitutions" for AI systems that lay out their fundamental purpose, principles, and constraints. These constitutions could be developed through a participatory process involving multiple stakeholders - from the AI developers themselves to domain experts, policymakers, and the general public.

The goal is to create a kind of social contract between humans and machines - one that ensures that AI systems are not just intelligent, but also ethical and trustworthy. And by codifying these principles into the very architecture of the systems we build, we can create a kind of moral scaffolding that guides their behavior and keeps them aligned with human values even as they become more and more autonomous.

Of course, none of this is going to be easy. As anyone who's ever tried to get a group of people to agree on a set of moral principles can attest, ethics and values are inherently messy and contested. And translating those fuzzy human concepts into the precise language of computer code is a challenge that's going to require deep interdisciplinary collaboration and a lot of hard work.

But I believe it's a challenge we have to take on if we want to create a future where artificial intelligence is a force for good in the world. We can't just cross our fingers and hope that these systems will somehow magically align themselves with human values - we need to be proactive in shaping them from the ground up.

And that's where you come in. As the next generation of AI researchers, developers, and leaders, you have an incredible opportunity to help steer the course of this technology and ensure that it benefits humanity as a whole. By grappling with these tough ethical questions now, and by advocating for approaches like value learning, inverse reinforcement learning, and constitutional AI, you can help to create a future where machines are not just intelligent, but also wise and benevolent.

It won't be easy, and there will undoubtedly be setbacks and challenges along the way. But I believe that if we approach this work with humility, care, and a deep commitment to human flourishing, we can create artificial intelligence that doesn't just serve us, but actually makes us better - as individuals, as societies, and as a species.

8.2. Responsible AI By Design

We've just covered the importance of aligning AI systems with human values and ethics and some of the approaches that researchers are exploring to try to make that happen. But if we really want to create AI that's truly responsible and trustworthy, we need to go beyond just thinking about the end goals or objectives of these systems.

Instead, we need to start building responsibility into the entire life-cycle of AI development - from the very first stages of problem formulation and data collection, all the way through to testing, deployment, and ongoing monitoring. It's about creating a culture and a process of ethical AI design that infuses these considerations into every step of the development process.

So what does that actually look like in practice? Well, one key element is making sure that we're not just optimizing for raw performance or efficiency, but also taking into account broader societal values and impacts. That means thinking carefully about things like fairness, accountability, transparency, and privacy from the very beginning, and making sure they're baked into the design of the system itself.

For example, let's say you're building an AI system to help make hiring decisions for a company. It's not enough to just focus on creating an algorithm that can accurately predict job performance based on resumes and interviews. You also need to think about potential biases in your training data, and how those might lead to unfair or discriminatory outcomes.

You need to consider the privacy implications of collecting and storing sensitive personal information about job candidates. You need to think about how you'll ensure that the system's decisions are transparent and explainable to the humans who are using it. And you need to have clear processes in place for accountability and redress if something does go wrong.

These are the kinds of considerations that need to be part and parcel of the AI development process from day one. And there are emerging frameworks and methodologies that can help guide this kind of responsible AI design.

One prominent example is the IEEE's Ethically Aligned Design framework, which lays out a set of principles and practices for embedding human values into autonomous and intelligent systems. The framework emphasizes things like human rights, well-being, accountability, transparency, and fairness as core pillars of ethical AI design.

Another key concept is the idea of "ethics by design" - essentially, the notion that ethical considerations should be built into the very architecture of AI systems from the ground up, rather than just being bolted on as an afterthought. This might involve things like incorporating explicit ethical constraints or objectives into the system's reward function or using techniques like inverse reward design to infer and align with human values.

But of course, even with the best of intentions and the most carefully designed frameworks, there's always the risk that AI systems might behave in unexpected or unintended ways once they're out in the real world. That's why another crucial element of responsible AI development is rigorous testing and validation to identify and mitigate potential risks and negative impacts.

One powerful technique that's gaining a lot of attention in this space is adversarial testing - essentially, trying to actively break or subvert the system to expose its weaknesses and blind spots. This might involve things like deliberately feeding the system biased or misleading data to see how it responds or using advanced optimization techniques to find edge cases where the system might behave badly.

Another approach is red-teaming - essentially, bringing in an outside group of experts to try to find flaws and vulnerabilities in the system that the original developers might have missed. This can be especially valuable for identifying potential negative impacts or unintended consequences that might not be immediately obvious.

And perhaps most importantly, responsible AI development means being proactive in anticipating and modeling the potential long-term impacts of these systems - not just in terms of their direct outputs or behaviors, but also in terms of how they might interact with and shape the broader social, economic, and political contexts in which they're deployed.

This is where techniques like causal modeling and simulation can be incredibly valuable - allowing us to explore different scenarios and "what if" questions to get a sense of how an AI system might ripple out and affect the world over time. And by identifying potential risks and negative impacts early on, we can start to develop strategies and interventions to mitigate them before they become major problems.

Now, I know all of this might sound like a lot of extra work and complexity on top of the already daunting task of building advanced AI systems. And it's true that doing responsible AI design right requires a significant investment of time, resources, and interdisciplinary expertise.

But I would argue that it's an investment we can't afford not to make. Because the stakes are simply too high to leave the development of these incredibly powerful technologies to chance or to let them be shaped solely by the imperatives of profit or efficiency.

We've seen what happens when we rush headlong into new technologies without fully considering their social and ethical implications. The history of the 20th century is littered with examples of innovations that had devastating unintended consequences - from the environmental impacts of industrialization to the societal upheavals of globalization and automation.

And with AI, the potential risks and negative impacts are arguably even greater, given the sheer scale and power of these systems. We're talking about technologies that could reshape entire industries, disrupt social and political institutions, and even challenge fundamental assumptions about human agency and identity.

So if we want to create a future where AI is a positive force for good in the world, we need to start taking these ethical considerations seriously from the very beginning. We need to build responsibility into the DNA of these systems and create a culture of AI development that prioritizes human values and long-term flourishing over short-term gains.

And that's a challenge that I believe falls squarely on the shoulders of your generation. As the creators and stewards of these technologies, you have an incredible opportunity - and an incredible responsibility - to shape them in a way that benefits humanity as a whole.

It won't be easy, and there will undoubtedly be difficult trade-offs and uncomfortable conversations along the way. But I believe that if we approach this work with care, humility, and a deep commitment to building a better world, we can create an AI future that we can all be proud of.

8.3. AI Governance And Oversight

We've talked about the importance of designing AI systems that are aligned with human values and that prioritize ethical considerations throughout the development process. But even with the most carefully designed systems, there's always going to be a need for ongoing governance and oversight to ensure that these technologies are being deployed and used in a responsible and accountable way.

And this is especially true as AI systems become more and more sophisticated and start being used in high-stakes domains like healthcare, finance, transportation, and criminal justice. The decisions that these systems make can have profound impacts on people's lives and livelihoods, and we need to have robust mechanisms in place to ensure that they're being used in a way that's fair, transparent, and in line with societal values.

So what might these governance mechanisms look like in practice? Well, one key idea that's gaining a lot of traction is the notion of AI ethics boards or oversight committees. Essentially, these are groups of experts and stakeholders from different domains - including technologists, ethicists, legal scholars, domain experts, and representatives of affected communities - who are tasked with auditing and monitoring the use of AI systems in a particular context.

These boards could be set up at the organizational level - for example, a company that's developing or deploying AI systems could create an internal ethics board to review and approve projects, sort of like how many universities have Institutional Review Boards (IRBs) to oversee research involving human subjects. Or they could be set up at the industry or sector level, to provide more broad-based oversight and governance across multiple organizations working on similar applications.

The key is that these boards need to have real teeth and real authority. They can't just be a rubber stamp or a PR exercise - they need to have the power to actually investigate and audit AI systems, to demand changes or modifications when necessary, and to hold developers and deployers accountable for any harm or negative impact.

And crucially, these boards need to be diverse and representative of the communities and stakeholders that are affected by the AI systems in question. We can't just have a bunch of Silicon Valley tech bros making decisions about how these technologies will impact people's lives - we need to have a range of perspectives and voices at the table, including those from marginalized or vulnerable groups who are often disproportionately affected by these systems.

But of course, ethics boards and oversight committees are just one piece of the governance puzzle. There's also a critical role for policy and regulation at the societal and governmental level to ensure that AI is being developed and deployed in a responsible and accountable way.

This could take a variety of forms, depending on the specific context and application. At the softer end of the spectrum, we might see things like voluntary industry standards or best practices that set out guidelines for ethical AI development and deployment. These could be developed through multi-stakeholder processes that bring together technologists, policymakers, civil society groups, and other key players to hash out shared principles and commitments.

There are already some promising examples of this kind of soft governance in action. For example, the IEEE's Ethically Aligned Design standards provide a set of principles and practices for embedding human values into autonomous and intelligent systems. And the Partnership on AI, a multistakeholder group that includes leading tech companies, academic institutions, and civil society organizations, has developed a set of tenets and best practices for responsible AI development.

But of course, voluntary standards and best practices can only take us so far. In many cases, there's going to be a need for harder forms of governance and regulation, with real legal force and consequences for non-compliance.

This might include things like mandatory testing and certification requirements for AI systems before they can be deployed in certain high-stakes domains. It could involve liability frameworks that hold developers and deployers accountable for any harm or negative impact caused by their systems. And it could include mechanisms for ongoing monitoring and auditing to ensure that AI systems are operating as intended and not causing unintended consequences.

We're already starting to see some early examples of this kind of hard regulation emerging in places like the European Union, which is currently developing a comprehensive legal framework for AI that includes mandatory risk assessments, transparency requirements, and human oversight provisions. And in the US, there are a number of bills and proposals circulating at the federal and state level that would start to put some regulatory guardrails around AI development and deployment.

But of course, getting the policy and regulatory balance right is going to be a huge challenge. We need to find ways to promote innovation and reap the benefits of these powerful technologies, while also ensuring that they're being developed and used in a way that's ethical, accountable, and in line with societal values. And we need to do it in a way that doesn't stifle creativity or bog down the development process in red tape.

It's a delicate balance, and one that's going to require a lot of ongoing dialogue, experimentation, and collaboration between technologists, policymakers, and other key stakeholders. And it's a balance that we're going to have to continue to adjust and adapt as the technology itself evolves and new challenges and opportunities emerge.

But I believe that getting this balance right is one of the most important challenges we face as a society in the coming decades. Because the stakes are simply too high to leave the governance of these incredibly powerful technologies to chance, or to the whims of the market.

We need to be proactive in shaping the development and deployment of AI in a way that maximizes its benefits and minimizes its risks. And that's going to require all of us - developers, policymakers, activists, and everyday citizens - to be engaged and involved in the process.

So I encourage all of you, as the next generation of leaders and innovators, to start grappling with these governance challenges now. Educate yourselves about the issues, engage in the ongoing debates and discussions, and lend your voices and your ideas to help shape the future of AI governance.

Because ultimately, the AI systems that we create today will shape the world that you inherit tomorrow. And it's up to all of us to work together to ensure that it's a world that we can be proud of - one that's more just, more equitable, and more in line with our deepest values and aspirations as human beings.

8.4. Ethical Deployment Of AI In Organizations

Now, let's talk about something that's on a lot of people's minds these days: the ethical deployment of AI in organizations, and how we can make sure it doesn't lead to massive layoffs and job losses. Because let's face it - as exciting as all this AI stuff is, it's also raising some pretty big questions and concerns about the future of work and the role of humans in a world increasingly dominated by machines.

But here's the thing - it doesn't have to be a zero-sum game. With the right strategies and approaches, organizations can actually leverage AI in a way that creates new opportunities, empowers employees, and minimizes job losses. And it all starts with one key ingredient: collaboration.

I'll repeat myself here, when it comes to deploying AI in an organization, it's not just about the technology. It's about the people - the stakeholders who will be impacted by that technology, from the frontline workers to the C-suite executives. And if you want to make sure that the deployment process is ethical, effective, and equitable, you need to bring all those stakeholders to the table.

That means involving employees and unions in the planning and decision-making process, not just as an afterthought, but as active participants. It means creating open lines of communication and transparency, so that everyone understands what's happening and why. And it means working together to identify potential challenges and opportunities, and to develop strategies that benefit everyone.

One of the key opportunities that AI presents is the chance to create new job roles and responsibilities within the organization. Think about it - as AI takes over certain tasks and functions, it's also creating a need for people to manage, monitor, and maintain those AI systems. It's creating a need for people with specialized skills and knowledge, who can help bridge the gap between the technology and the business.

So instead of just focusing on the jobs that might be lost to AI, organizations need to be proactive about identifying and creating new job opportunities that leverage AI. That might mean retraining and redeploying employees whose current roles are impacted by AI, so that they can take on new responsibilities and add value in different ways. It might mean investing in education and skill-building programs, to help employees stay ahead of the curve and adapt to changing job requirements.

And here's the best part - there are already plenty of examples out there of organizations that are doing this successfully. Take Siemens, for example - the German industrial giant has been investing heavily in AI and automation in recent years, but instead of laying off workers, they've been retraining them to work alongside the new technology. They've created new roles like "robot coordinator" and "data scientist," and have even set up their own in-house training programs to help employees acquire the skills they need to thrive in a more automated workplace.

Or look at Amazon - the e-commerce behemoth has been one of the biggest adopters of AI and automation in recent years, using everything from robots in its warehouses to algorithms that recommend products to customers. But even as it has automated many tasks and functions, Amazon has also been investing in its workforce, offering training and education programs to help employees move into higher-skill, higher-paying roles within the company.

Of course, these examples aren't perfect - there are still plenty of challenges and concerns around the impact of AI on jobs and workers. But they show that it is possible to deploy AI in a way that is ethical and employee-friendly, if organizations are willing to put in the work and collaborate with their stakeholders.

So what can we learn from these examples? Well, for one thing, it's clear that communication and transparency are key. Employees need to be kept in the loop about what's happening with AI in their organization, and they need to have a voice in the decision-making process. That means regular updates, town hall meetings, and opportunities for feedback and input.

It's also clear that investing in training and education is crucial. As AI changes the nature of work, employees need to be equipped with the skills and knowledge to adapt and succeed in new roles. That means providing access to training programs, educational resources, and career development opportunities.

And perhaps most importantly, it means approaching AI deployment with a human-centered mindset. It means recognizing that the ultimate goal is not just to maximize efficiency or cut costs, but to create a workplace that is fair, equitable, and empowering for everyone involved.

Because at the end of the day, AI is just a tool - it's the people who use that tool, and the intentions behind it, that really matter. And if we can approach AI deployment with empathy, collaboration, and a commitment to ethical principles, we can create a future of work that benefits everyone - not just the machines, but the humans who work alongside them.

So let this be a call to action - for organizations, for employees, and for anyone who cares about the future of work in the age of AI. Let's come together, let's have the hard conversations, and let's find a way to make this technology work for us, not against us.

8.5. Anthropic AI And Human Flourishing

Alright, folks, it's time to get a little bit sci-fi and a whole lot visionary as we close out this chapter on ethical AI development. Because while a lot of the conversations around AI ethics tend to focus on the potential risks and downsides of these technologies - and don't get me wrong, those are absolutely critical considerations - I think it's just as important that we spend some time dreaming big about the incredible positive potential of AI to help humanity flourish.

And that's really what the emerging field of "anthropic AI" is all about. This is the idea of developing AI systems that are not just designed to avoid harm or to be neutral in their impacts, but that are actively and intentionally designed to promote human well-being and flourishing in all its forms.

It's a vision of AI that's less about creating superhuman intelligences that can beat us at chess or write poetry, and more about creating tools and partners that can help us be the best versions of ourselves - that can enhance our creativity, expand our knowledge and capabilities, and help us solve some of the biggest challenges we face as a species.

So what might this kind of "anthropic AI" actually look like in practice? Well, one key area of focus is on using AI to enhance and augment human cognitive capabilities. We're already starting to see early examples of this with things like AI-powered tutoring systems that can adapt to each learner's individual needs and learning styles, or with tools like GPT-4 that can help writers and creatives generate new ideas and content.

But the potential here goes far beyond just making us a little bit smarter or more creative. Imagine AI systems that could help us navigate the vast amounts of information and knowledge that are being generated every day, and surface the most relevant and valuable insights for our needs. Imagine AI that could help us make better decisions by taking into account far more variables and consequences than our human brains can handle. Imagine AI that could even help us understand and communicate with each other better, by bridging linguistic and cultural divides and fostering greater empathy and understanding.

Another key area where anthropic AI could have a transformative impact is in the realm of scientific discovery and technological innovation. We're already seeing incredible examples of AI being used to accelerate breakthroughs in fields like drug discovery, materials science, and renewable energy. By crunching through vast amounts of data and running millions of simulations, AI is helping researchers identify promising new candidates for everything from cancer treatments to battery technologies in a fraction of the time it would take human scientists alone.

But again, the potential here is almost limitless. Imagine AI systems that could help us design new sustainable cities from the ground up, optimizing for things like energy efficiency, walkability, and social cohesion. Imagine AI that could help us create new forms of art and music that push the boundaries of human creativity and expression. Imagine AI that could even help us unlock the secrets of the universe itself, by aiding in the search for new fundamental particles or the detection of gravitational waves.

And perhaps most excitingly of all, anthropic AI could be a powerful tool in helping us tackle some of the most pressing and existential challenges facing humanity today. Things like climate change, global poverty and inequality, pandemic preparedness, and even the specter of aging and age-related disease.

These are challenges that are simply too big and too complex for any one human mind or even any one country or organization to solve alone. They require a level of coordination, creativity, and sheer brainpower that we've never had access to before. But with the help of AI systems that are designed from the ground up to promote human flourishing, we may finally have a chance at cracking some of these "grand challenges" once and for all.

Imagine AI that could help us develop new clean energy technologies and rapidly transition away from fossil fuels. Imagine AI that could help us create more efficient and equitable systems for distributing food, healthcare, and other basic necessities to people in need around the world. Imagine AI that could even help us crack the code of biological aging, allowing us to live longer, healthier lives and spend more time doing the things we love with the people we love.

Now, I know that some of these visions might sound like pure science fiction, or even like techno-utopian fantasies. And it's true that we still have a long way to go before we can realize the full potential of anthropic AI. There are technical challenges to overcome, ethical considerations to grapple with, and societal implications to navigate.

But I believe that this is a vision worth pursuing, and one that we can't afford to shy away from. Because at the end of the day, the ultimate goal of all our technological progress and innovation should be to make life better for everyone on this planet. To reduce suffering, to expand opportunity, to push the boundaries of what's possible.

And if we can create AI systems that are truly in service of those goals - that are designed from the ground up to promote human flourishing in all its forms - then I believe that we'll look back on this moment as a turning point in human history. A moment when we finally harnessed the power of our technology to create a world that's more just, more equitable, and more in line with our deepest values and aspirations.

It won't be easy, and there will undoubtedly be challenges and setbacks along the way. But I believe that this is a challenge that we must take on, and one that your generation in particular is uniquely positioned to lead us through.

So as you go out into the world and start grappling with these big questions and challenges, I encourage you to keep this vision of anthropic AI in mind, to ask yourselves not just how we can create AI that's safe and ethical, but how we can create AI that truly helps us flourish as individuals and as a species.

Because if we can get this right - if we can create a future where AI is a powerful partner in our quest to create a better world - then I believe that there's no limit to what we can achieve together. Personally, I like to stay optimistic about the future. And I can't wait to see what it holds for us all.

CHAPTER 9

CONCLUSION: EMBRACING OUR COLLABORATIVE FUTURE WITH AI

We've covered a lot in this book, haven't we? From the fundamentals of AI and machine learning to the incredible ways that these technologies are transforming industries and skills, to the importance of cultivating our own unique human strengths in the face of automation and change. It's been quite the journey!

But as we come to the end of this exploration, I want to leave you with one final message - and that's a message of hope, excitement, and possibility. Because while it's true that the age of AI is going to bring some big challenges and disruptions to the way we live and work, it's also going to bring some truly incredible opportunities for growth, innovation, and human flourishing.

You see, despite all the hype and fear-mongering out there, the rise of artificial intelligence is not about machines replacing humans, or about some dystopian future where we're all rendered obsolete. It's about creating a new kind of partnership between humans and technology - one where we work together in ways that enhance and amplify our unique strengths and capabilities.

Think about some of the examples we've explored throughout this book. Like the doctors who are using AI tools to help them diagnose diseases and develop personalized treatment plans faster and more accurately than ever before. Or the creative professionals who are using machine learning algorithms to generate new ideas and insights, and to push the boundaries of what's possible in fields like art, music, and design.

In each of these cases, the goal is not to replace human intelligence and creativity with machine intelligence - it's to augment and enhance it. It's about creating a world where humans and AI can work together in a collaborative, symbiotic way, each bringing their own unique strengths and abilities to the table.

Let's take a look at one more example - this book. I wrote it in collaboration with AI, Anthropic's Claude to be exact. Throughout the writing process, Claude and I engaged in a creative partnership, with the AI contributing ideas, examples, and prose that I curated, edited, and stitched together with my own insights and narrative vision. The result is a uniquely blended work that leverages both human and machine intelligence in service of a shared goal. It's a powerful illustration of the kind of human-AI collaboration that I believe will become increasingly common and impactful in the years ahead.

Now, my dear reader, as the next generation of leaders, innovators, and changemakers, you have an incredible opportunity to shape the future of this human-AI collaboration. You have the chance to bring your own distinct blend of creativity, empathy, and strategic thinking to the table, and to help create a world where technology is a tool for empowerment and liberation, rather than a force for displacement and despair.

So as you go out into the world and start navigating your own path in the age of AI, I want you to remember a few key things.

First and foremost, remember that your human skills and abilities are your ultimate competitive advantage. In a world where more and more tasks and jobs are being automated, it's the things that make us distinctly human - our creativity, our emotional intelligence, our ability to communicate and collaborate with others - that will truly set us apart. So don't be afraid to prioritize and cultivate those skills, even as you're also learning about and leveraging the power of AI.

Second, remember that success in the age of AI is not about fighting against the machines, but about learning to work with them in a way that maximizes your own unique value proposition. Whether that means using AI tools to automate routine tasks and free up your time for more creative and strategic work, or collaborating with machine learning algorithms to generate new insights and ideas, the key is to approach technology as a partner and an enabler, rather than a threat.

And finally, remember that the future is not some fixed destination that we're all just hurtling towards - it's something that we have the power to shape and create ourselves. So don't be afraid to dream big, to take risks, and to use your own unique blend of human and technological skills to tackle the biggest challenges and opportunities of our time. Whether that's solving climate change, curing disease, or creating entirely new industries and ways of living and working, the possibilities are truly endless.

Of course, I won't pretend that any of this will be easy. The age of AI is going to bring plenty of challenges and uncertainties along the way, and there will undoubtedly be times when it feels like the machines are getting the upper hand. But if there's one thing that I've learned in all my years of studying and working with technology, it's that the human spirit is incredibly resilient and adaptable. We have a remarkable capacity for creativity, innovation, and collaboration - and when we put our minds and our hearts into something, there's very little that we can't achieve.

So as you go out into the brave new world of AI and automation, I want you to hold onto that spirit of resilience and possibility. I want you to remember that you have unique gifts and talents that are valuable and needed in this world, and that by bringing those to the table and collaborating with others - both human and machine - you have the power to create truly incredible things.

The future may be uncertain, but one thing I know for sure is that it's going to be an amazing ride. And I can't wait to see all the incredible ways that you'll use your own human edge to shape it for the better. So go out there and make your mark - the world is waiting for you!

ABOUT THE AUTHOR

Roman Kurmachev is a passionate AI enthusiast with a background in Linguistics and years of experience working in the tech industry in Silicon Valley. He has been closely following the developments in Natural Language Processing (NLP) and has witnessed firsthand the transformative power of AI in the workplace.

Through his own experiences and extensive research, Roman has developed a unique perspective on the future of work and the role of humans in an AI-driven world. He is passionate about sharing his insights and strategies for thriving in the age of AI and empowering others to embrace the opportunities and challenges of this transformative technology.

With "Outsmart the Algorithms: Thriving in the Age of AI," Roman aims to provide a comprehensive and accessible guide for navigating the complexities of human-AI collaboration, and inspiring readers to harness their uniquely human skills and qualities to create a brighter, more hopeful future.

www.ingramcontent.com/pod-product-compliance
Lightning Source LLC
Chambersburg PA
CBHW060537130626
46553CB00002B/799